Challenges for European Innovation Policy

Cohesion and Excellence from a Schumpeterian Perspective

Edited by

Slavo Radosevic

Professor of Industry and Innovation Studies, School of Slavonic and East European Studies, University College London, UK

Anna Kaderabkova

Director of Centre for Economic Studies in Prague, Czech Republic

Edward Elgar
Cheltenham, UK • Northampton, MA, USA

Published by
Edward Elgar Publishing Limited
The Lypiatts
15 Lansdown Road
Cheltenham
Glos GL50 2JA
UK

Edward Elgar Publishing, Inc.
William Pratt House
9 Dewey Court
Northampton
Massachusetts 01060
USA

A catalogue record for this book
is available from the British Library

Library of Congress Control Number: 2011925704

ISBN 978 1 84980 309 0 (cased)

Typeset by Servis Filmsetting Ltd, Stockport, Cheshire
Printed and bound by MPG Books Group, UK

Contents

Figures

Tables

Boxes

Contributors

Philippe Aghion, Professor of Economics, Harvard University, USA

Jakob Edler, Professor of Innovation Policy and Strategy and Executive Director of the Manchester Institute of Innovation Research, Manchester Business School, University of Manchester, UK

Heike Harmgart, Principal Economist, European Bank for Reconstruction and Development, London, UK

Anna Kaderabkova, Director of the Centre for Economic Studies, Prague, Czech Republic

Rajneesh Narula, Professor of International Business Regulation and Director of the John H. Dunning Centre for International Business, University of Reading, UK

Slavo Radosevic, Professor of Industry and Innovation Studies, School of Slavonic and East European Studies, University College London, UK

Alasdair Reid, Director of Technopolis, Belgium

Andreas Reinstaller, Director of Technopolis Group, Belgium

Fabian Unterlass, Austrian Institute for Economic Research (WIFO), Austria

Natalia Weisshaar, Department of Economics, Royal Holloway, University of London, UK

Abbreviations

BERD	business expenditure on research and development
CEB	Central Eastern Europe and the Baltic States
CEE	Central and Eastern Europe
CEEC5	The five Central and Eastern European Countries (Czech Republic, Hungary, Poland, Slovakia and Slovenia)
CIS*	Commonwealth of Independent States
CIS+M	Commonwealth of Independent States and Mongolia
CIS*	Community Innovation Survey
DBIP	demand-based innovation policies
EBRD	European Bank for Reconstruction and Development
EC	European Commission
EIS	European Innovation Scoreboard
EPO	European Patent Office
EU	European Union
FDI	foreign direct investment
FP	Framework Programme(s)
GDP	Gross Domestic Product
GFC	global financial crisis
ICT	Information and Communication Technology
IPR	Intellectual Property Rights
KIS	knowledge-intensive services
LIUP	Local Industry Upgrading Programme
MNC	multinational corporation(s)
MNE	multinational enterprise(s)
NIC	national innovation capabilities
NIS	national innovation system(s)
NMS	New Member States
NSI	national system(s) of innovation
OECD	Organisation for Economic Cooperation and Development
OMC	open method of coordination
PISA	Programme for International Student Assessment
PPS	purchasing power standard
R&D	research and development
RTD	research, technology and development
RTDI	research, technological development and innovation

SEE	South Eastern Europe
SF	Structural Funds
SII	summary innovation index
SME	small and medium sized enterprises
STI	science, technology and innovation
STIG	science, technology, innovation and growth
TFP	total factor productivity
UNIDO	United Nations Industrial Development Organization
US	United States (of America)
WEF	World Economic Forum
WTO	World Trade Organization

*Note: We use CIS to mean Community Innovation Survey and also Commonwealth of Independent States. Their different usage is clear in the chapters where they occur.

1. Innovation policy in multi-tier Europe: introduction

Anna Kaderabkova and Slavo Radosevic

This volume is the result of an expert group assembled under the auspices of the Czech European Union (EU) presidency in 2009. The group was mandated to explore innovation policy issues in the EU in light of the increased divergence of the EU27 and the realization that these differences require different policy approaches. The enlarged EU27 represents a new architecture of countries that can no longer be defined by a simple East–West divide, with most innovation indicators suggesting the existence of a multi-tier Europe in terms of innovation capacity (Radosevic, 2004; EC, 2009; also see Chapter 8 in this volume, Figure 8.1). It would be unrealistic, therefore, to expect that similar policies and indicators could be used to gauge and benchmark the innovation performance of these countries. Rather than a criterion of frontier technology, policies need to encompass sets of criteria relevant to the technology profiles of individual countries. This calls for a perspective that takes account of *the cross-country heterogeneity of existing national systems of innovation (NSI) and the technology positions of countries in relation to the technology frontier*. A country-specific approach should be combined with sector-specific approaches, which highlight the *diversity of technology modes* across industry. This understanding of innovation and growth in Europe is not compatible with current policy practice. The main tenets of EU-level innovation policy were conceived before accession of the new member states (NMS) and were based on generic and horizontally-oriented policies applied in a uniform manner across a large number of countries. These would seem to be insufficient since they involve long gestation and do not cater to the specific requirements of groups of countries based on their different distance from technological frontier countries.

This thinking is based on the logic underpinning (neo)Schumpeterian growth theory (Aghion and Howitt, 1998), the perspective used in this volume to explore innovation policy issues. (Neo)Schumpeterian theory implies that policies and institutions can have long-lasting effects on economic growth, and that they should differ depending on the position of a country in relation to the technology frontier. This book provides a unique application of the Schumpeterian innovation policy perspective in the context of the countries of Central and Eastern Europe (CEE). It is an example of the kind of applied work formally developed by Philippe Aghion, a contributor to this volume, and by Richard Nelson, Chris Freeman, Carlota Perez, Luc Soete, Bengt-Ake Lundvall, Franco Malerba and colleagues, in a range of neo-Schumpeterian 'appreciative theorizing' type contributions. What distinguishes this book is that it employs this perspective explicitly and systematically, in the context of the EU, and EU new member CEE states in particular. The approach is applied on several dimensions (sectoral, demand, foreign direct investment (FDI), finance, education and innovation policy proper), combining empirical and conceptual analysis.

The focus on EU and the NMS in this book is logical given the multi-tier European context in which CEE countries are operating as peripheral economies in terms of technology generation. It stems also from an understanding that the same policies may not be effective for countries at very different distances from the world technology frontier. As Aghion et al. (2006: 13) point out, growth policy requires that the 'characteristics of the country or sector such as its degree of technological development or the extent of financial constraints, the nature of slow moving institutions in that particular country or sector, be taken into account'.

Projected growth in the EU27 is based on innovation and knowledge – which translates into different sets of requirements for different countries. Nevertheless, in all cases growth requires *complementary policies*, that is, innovation policy and structural reform that is coherent with competition, higher education, labour market flexibility and financial market development policies, all of which must be tailored to the individual countries of the EU27. For example, an exclusive focus on entry and exit barriers in new technology intensive sectors is relevant for industries and countries close to the technology frontier, but perhaps not for those that are farther away. As has been demonstrated elsewhere (Arundel et al., 2008), there has been a general failure in policy to support non-research and development (R&D) innovators to upgrade their innovative capabilities. This applies particularly to CEE, where the share of R&D innovators is disproportionately high, but policy is focused on support for R&D innovation. There also needs to be good coherence between structural and macroeconomic policies throughout the business cycle (Aghion, 2006). This is currently lacking in Europe, and CEE countries are increasingly suffering the effects of unaligned policies. As Aghion (2006) argues, countercyclical budgetary policies are more growth-enhancing in countries with less developed financial systems and, thus, their absence in CEE, whose financial development is poor, is very detrimental.

The general focus of this volume is on innovation policy, but within a broadly defined framework of science, technology, innovation and growth (STIG) systems (Aghion et al., 2006). Specifically, we examine the (in) capacity of CEE to play a more important role in the knowledge-based competitiveness of the EU. We explore whether it is possible to bolster this capacity with innovation/technology/industry-specific policies, and discuss the changes that would be needed at EU and individual country levels to remove sector (industry)-specific obstacles to greater competitiveness based on innovation. We analyse these policies from the perspective of growth, which makes their investigation informative for education, labour market and competition policy.

In countries and industries that are far from the technology frontier, technology transfer and non-R&D-based innovation activities are important drivers of innovation. In these countries, increasing levels of technology transfer and absorptive capacity through R&D and training should be a priority. In more advanced countries which rely on knowledge-intensive foreign imports, innovation policy should support a gradual move from the acquisition to the production of knowledge. Countries at or close to the technology frontier should foster R&D in companies and universities. In sum, innovation policy needs to be country-specific in the sense of tackling those aspects of national innovation capacity that are major bottlenecks.

The contributors to this volume are critical of EU-wide policy approaches that ignore individual country needs. The use of a single yardstick, whether policy metrics such as the European Innovation Scoreboard, methods such as the open method of coordination, or imposed policy objectives such as the 3 per cent, take no account of these differences. Policies for countries where growth is based on technology absorption need to be different from policies applied to countries where growth is based on R&D and technology frontier innovation.

Nevertheless, there are some innovation challenges that apply to all countries and sectors such as shortages of skilled human resources and financing for business R&D. The policy mix needs to be broad-based and to take account of 'framework conditions'. Only specific or only general policies will not be effective and are not substitutes for one another.

Such issues are the core of this volume and the basis of its coherence. We believe it will be essential reading for those interested in the issue of innovation policy in Europe.

Chapter 2 by Slavo Radosevic provides an analytical introduction to and context for the succeeding chapters. It positions innovation and innovation policy issues in CEE countries within the enlarged EU, and draws comparisons with the developed EU. It outlines the rationale for developing a new policy approach, endorsed by the other contributors to this volume, which takes account of technological and developmental differences across countries. It outlines some policy implications of a specific country/country group approach which are developed in the following chapters.

Chapter 3 by Philippe Aghion, Heike Harmgart and Natalia Weisshaar introduces a general framework for designing medium term growth-enhancing policies through a focus on specific policies relating to competition, education and finance. It reviews the growth experiences of the European transition countries and argues that general and non-sector-specific government intervention can substantially increase long-term growth prospects in these countries. While specific sectors might receive greater benefits from specific policy measures, this chapter emphasizes the contribution of overarching growth-enhancing policies. It focuses on two areas where policy could be particularly effective, namely competition and education quality. First, if transition countries are to achieve – and sustain – higher long-run growth rates, they will need to support competition by continuing to remove entry and trade barriers and by strengthening – and in some cases setting up – competition agencies. This applies particularly to the Commonwealth of Independent States (CIS), resource-rich countries. Second, the transition countries as a group need to invest more in the quality of primary and secondary education, which, in turn, implies that they need to invest more in tertiary, especially undergraduate, education to produce better quality teachers, and in the evaluation and monitoring of their education systems. Again, somewhat paradoxically, it is the resource-rich CIS countries that are suffering most from insufficient investment in education and problems related to the quality of their education services. This chapter argues finally that there may be scope for macroeconomic policies to boost spending in these key areas. In the area of education, in particular, the role of the private sector in overcoming skills mismatches will benefit from more intensive financial intermediation and a reduction in the barriers to finance.

Chapter 4 by Andreas Reinstaller and Fabian Unterlass explores sectoral innovation policy and its relevance and implications for the NMS from CEE. First, they establish the extent to which industry specialization affects the assessment of national innovation capability. They develop a country classification based on the technological profiles of EU countries and discuss the construction of meaningful industry classifications, based on the existing literature. The aim is to reduce the heterogeneity of firm innovation behaviour across industries and countries to a few salient types, enabling meaningful policy conclusions. Second, the authors analyse how industries differ in terms of their innovation strategies and how they exploit innovation opportunities, among the country groups identified in the first part of the chapter. The authors conclude with a discussion of how a better understanding of national specialization profiles and sectoral innovation modes could contribute to the design of policy mixes that would foster innovation, competitiveness and growth.

Chapter 5 by Alasdair Reid analyses the extent to which innovation policies in the CEE NMS reflect their individual industry and technology conditions and levels. This chapter examines the challenges related to designing innovation policies tailored to these states. It discusses the importance of types of 'system failure' in the NSI of the CEE countries and how they are being addressed by national policy; it reviews the innovation policy mix, including the role of EU Structural Funds, and its correspondence with the system failures identified. It argues in particular that CEE countries have specific innovation strengths and weaknesses, which call for the development of customized policies and not the blinkered application of a 'Europeanized' policy approach. The policy mix required for the CEE countries differs from policies appropriate for the 'average' EU25; in a number of cases the focus must be on resolving individual issues related to the stage of innovation (for example investment-driven approach to research, technology, development and innovation policy via technology acquisition by small and medium sized enterprises or promotion of the research infrastructure), tackling explicit bottlenecks in the NSI (for example doctoral studies in science and engineering, low levels of capabilities and spending on R&D by domestic enterprises). Reid concludes that there is a need to widen the scope of innovation policy, to exploit linkages with other policies aimed at education and training, rather than trying simply to remedy the perceived underinvestment in business R&D.

Chapter 6 by Rajneesh Narula explores the policy options for the integration of FDI and NSI, drawing on experience from around the world, but with a particular focus on EU NMS. He argues that the promotion of innovation activities by multinational enterprises (MNE) requires a somewhat different approach from the promotion of general value-adding activities. MNE innovative activities are increasingly fragmented across different locations to exploit particular aspects of different systems. Narula makes the distinction between two types of MNE R&D. Demand-driven R&D is activity undertaken to adapt existing products and services to local needs. Supply-side R&D is innovation conducted in independent, knowledge-intensive, R&D facilities and implies greater dependence on domestic knowledge sources and infrastructures. The two types of R&D also imply different technology and innovation policy options and require a focus on scope and competence at the MNE subsidiary level. Approaches that focus on FDI flows as aggregates are flawed, since knowledge exchange and innovation are establishment-level phenomena. The different focus advocated in this chapter requires the introduction of an MNE policy to link FDI and industrial policy. CEE NMS should focus on attracting and fostering demand-driven MNE R&D activities. Governments should reduce the emphasis on cost and increase the focus on specialized, location-bound knowledge assets, setting up programmes that will foster demand-oriented upgrading of public R&D and human capital.

Chapter 7 by Jakob Edler examines demand and other conditions required for innovation and demand-based policies for innovation, in the context of EU CEE countries (CEECs). His argument is that institutional adaptations, and a policy mix that tackles bottlenecks in demand for innovation and supports the articulation of demand, can link modernization of the economy and public services to increased innovation in the EU CEEC economies, thereby contributing to an identifiable catching up process. Chapter 7 introduces the wider context of current demand-based innovation policies at EU level; it provides a conceptualization of demand-based innovation policy and introduces a typology of demand measures. Edler argues that our understanding of EU CEE countries' innovation capabilities falls short when it comes to capturing the demand side. He discusses alternative variables to map demand conditions in the EU CEE countries, linking this discussion to the challenges for policy. The overall finding from this chapter is that, in the CEE countries, efforts – both public and private – to articulate and support demand for innovation, are not keeping pace with attempts to boost the supply (of science and technology) and to compete in the international race for better supply conditions. The task is not so much about the construction of a technocratic machine able to pick winners and steer markets, but rather is about the innovation culture, and improving the responsiveness of the NSI to newly emerging needs and challenges.

Chapter 8 is co-authored by all the contributors to this volume and develops policy implications from their analyses. We adopt a perspective that takes account of the cross-country heterogeneity of existing NSI and their positioning in terms of innovation capacity, and that also recognizes the need for country-specific 'policy mixes' and complementarities between innovation policy, structural reforms and other policies. We discuss existing policies and the need for a rethinking of the EU innovation agenda. For example, even though EU policies allow for flexibility in implementing the Lisbon Agenda, many countries are focusing on improvements in a few R&D-biased indicators (for example the Barcelona goal of 3 per cent of gross domestic product invested in R&D) and are ignoring the need for policy measures to foster national innovation capability and competitiveness. The Lisbon Agenda and other goals create strong public and peer pressure for improvements in some areas, which, in the case of some countries, could be detrimental since formal R&D is not the main innovation input for all industries and is less relevant during the process of catching up. This chapter highlights areas and actions that should be the focus of policy making in the EU, and especially the NMS. It discusses:

- the need to reconsider the balance between funding excellence and R&D relevance;
- the importance of improving education;
- provision of training;
- evaluating the use and design of Structural Funds to improve innovation capacity;
- institutional approaches for integrating FDI and innovation policy;
- spurring innovation through demand policies;
- investing in strategic intelligence and administrative capacity for more effective policy making.

The contributors to this volume constitute a team of people whose theoretical and empirical understanding of innovation policy issues in Europe is well recognized. The critical mass of these contributions and their clear focus on broad innovation policy in the enlarged EU makes this volume important reading for all those with an interest in these topics.

Innovation policy has become a mainstream activity. The innovation policy community has expanded much beyond the confines of government agencies, to encompass a broad array of consultancy firms, research groups, think tanks, students and enlightened public and business communities. We hope that this broad range of audiences will find this book interesting and informative and provide a better understanding of the issues of innovation and growth in the enlarged EU.

REFERENCES

Aghion, P. (2006), 'A primer on innovation and growth', *Bruegel Policy Brief*, 2006/06, Brussels: Bruegel.
Aghion, P. and P. Howitt (1998), *Endogenous Growth*, Cambridge, MA: MIT Press.
Aghion, P., P.A. David and D. Foray (2006), 'Linking policy research and practice in "STIG Systems": many obstacles, but some ways forward', *SIEPR Discussion Paper* No. 0609, Stanford, CA: Stanford Institute for Economic Policy Research.
Arundel, A., C. Bordoy and M. Kanerva (2008), 'Neglected innovators: how do

innovative firms that do not perform R&D innovate? Results of an analysis of the Innobarometer 2007 survey No. 215', *INNO-Metrics Thematic Paper*, Maastricht: MERIT.

European Commission (EC) (2009), *European Innovation Scoreboard*, Luxembourg: European Commission.

Radosevic, S. (2004), 'A two-tier or multi-tier Europe? Assessing the innovation capacities of central and east European countries in the enlarged EU', *Journal of Common Market Studies*, **42**(3), 641–66.

2. Challenges of converging innovation policies in a multi-tier Europe: a neo-Schumpeterian perspective[1]

Slavo Radosevic

2.1 INTRODUCTION

This chapter provides an introduction to and context for the other contributions in this volume. It discusses innovation and innovation policy in Central and East European (CEE) countries within the enlarged European Union (EU) and draws comparisons with the developed EU. It outlines the rationale for developing a new policy approach that takes account of technological and developmental differences across countries. It discusses the major issues in the approach to innovation policy of a specific country or group of countries.

From the end of the 1990s to the time of the global financial crisis (GFC) in 2008–09, growth in the CEE EU new member states (NMS) has been significantly above the rates of growth in the EU15.[2] This trend, coupled with the inflow of EU Structural Funds (SF), has led to a widespread belief that these countries are established on a path of catch-up and fast convergence to EU15 income levels. However, the GFC has left the NMS, and especially those with significant macroeconomic imbalances, vulnerable to the vagaries of the financial markets and has exposed the less than firm foundations of their growth, showing them to be inadequate to sustain growth in the longer term (World Bank, 2010).

Since the 2004 and 2006 accessions to the EU, growth in the CEE countries has been based on imported capital equipment and transition related productivity improvements, achieved through organizational changes and reallocation of the factors of production (EC, 2009a). Demand-side growth has been based on a combination of domestic demand and export activity. However, with the exception of the Czech Republic after 2005, and Hungary after 2008, net export levels have not been high. A process of real catch-up by the NMS will require that these countries continue to increase productivity growth rates, which will be dependent on their capacity to upgrade technologically. Technological upgrading involves both supply-side (research and development – R&D) improvements, and higher quality human capital (absorptive capacity), demand and diffusion (Radosevic, 2004b). It is expected that improvements in these four dimensions will be based on SF allocations, which, in many ways, have become the key to modernization for many countries. SF are providing a unique 'window of opportunity' for NMS, although historical experience would suggest that their contribution is far from sufficient for economic catching up (Barca, 2009).

We examine this issue from a Schumpeterian growth perspective (Aghion and Howitt, 1997). Schumpeterian growth theory explains sustained long-run growth by endogenizing productivity growth resulting from innovation with new profitable innovators successively displacing earlier innovators. Faster growth generally implies higher firm turnover through a process of creative destruction, which generates entry of new innovators and exit of former innovators. Schumpeterian growth theory introduces an important distinction between innovation and imitation. Imitator countries catch up to the global technology frontier, represented by the stock of global technological knowledge available to all innovators. A country's growth path and its innovation pattern are determined endogenously by the process of competition between a prospective innovator and the competing fringe of producers. From this perspective, the intensity and mix of innovation often depends upon institutions and policies, but in ways that vary with the distance of the particular country from the technological frontier (Aghion and Howitt, 2006). Accordingly, policies that are conducive to growth by a country's innovators are not necessarily favourable to growth by imitators, or by countries that are farther away from the technology frontier, which maximize growth by favouring institutions that facilitate imitation. CEE countries are behind the technology frontier, but some are working towards innovation-based growth (WEF, 2009). From a Schumpeterian perspective, their institutional frameworks and policies can be described as being in transition from imitation-enhancing to innovation-enhancing, as the relative importance of innovation for their growth increases (see Chapter 3 in this volume). In a nutshell, this perspective assumes that growth policies must be country-specific and based on the distance from the technology frontier (Aghion and Howitt, 2006).

This perspective is extremely relevant in the context of the EU27 countries where we observe: (i) diversity among countries in relation to distance to the technology frontier; and (ii) a widespread belief that convergence in growth levels requires convergence in policies. The latter aspect is observable in the institutional requirements for EU accession, including preferences for Eurozone membership, and innovation policy convergence based on various methods of open coordination and diffusion of best practice (Kaiser and Prange, 2004). The contributors to this volume express strong reservations about a convergent policy approach and propose some alternative policy thinking.

The focus of this chapter is innovation policy within a broadly defined framework of 'science, technology, innovation and growth systems' (STIGS) (Aghion et al., 2006). It focuses specifically on the issues and policies that affect the capacities of new members to play prominent roles in the knowledge-based competitiveness of the EU. We explore the ways that innovation policies in the NMS should differ from common EU-wide policy approaches. In Section 2.2 we discuss the main motivations for this chapter. Section 2.3 discusses the approach subscribed to by the authors of the chapters in this volume. Section 2.4 highlights some emerging analytical policy issues, and Section 2.5 concludes by summarizing some key points.

This edited book tries to advance policy thinking by framing the innovation policy agenda within a (neo)Schumpeterian perspective. Its point of departure is that innovation and industry policies are very dependent on a country's technology position in relation to the technology frontier, and an appropriate institutional context (Rodrik, 2008; Radosevic, 2009a). Locally specific constraints and opportunities determine the effectiveness of similar policies. This local 'know-how' is more important than which policies are implemented. All policy advice that takes account of the specific policy context will be relevant. We do not pretend to be able to offer ready-made solutions and neat answers. However, we believe that applying a Schumpeterian perspective is a useful and novel way to consider innovation policy in the context of the EU27.

2.2 NMS IN THE ENLARGED EU: HOW DIVERSE IS THE EU27?

Between the end of the 1990s and the GFC, following transition-related recession and a significant falling behind, the CEE countries managed to reduce their income and productivity gap with the developed EU (see Figures 2.1 and 2.2), which narrowed from 40 per cent to 51.7 per cent of EU15 gross domestic product (GDP) per capita (EC, 2009a: 32). However, the GFC revealed the fragility of their growth, indicating that it must be considered a spurt rather than technology-based catch-up. Growth spurts are periods of medium-term high growth where countries grow at much above historical rates, but at rates that are not sustainable in the long term (Hausman et al., 2004).

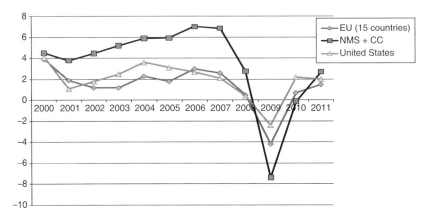

Source: Eurostat Structural Indicators.

Figure 2.1   Real GDP growth rates: CEE economies' spurt in the 2000s

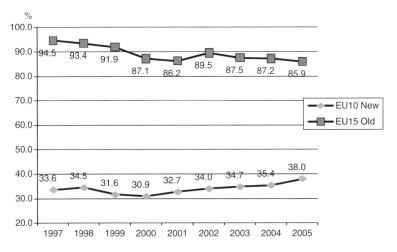

Source: Eurostat Structural Indicators.

Figure 2.2   Labour productivity per hour worked (US=100)

Technology-based catch-up is observed based on a country's position in relation to the current world technology leader, the US. Measured in terms of labour productivity, both the 'old' EU15 and the 'new' CEE EU10 are trailing, at 86 per cent and 38 per cent respectively of US productivity levels (Figure 2.2). Within this scenario, the EU15 has lost ground while the CEE NMS have narrowed the gap with the US. This temporally short

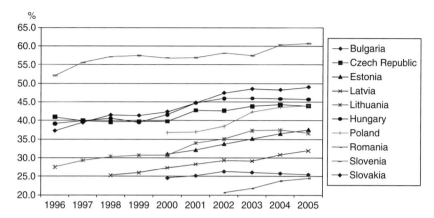

Source: Eurostat Structural Indicators.

Figure 2.3   Labour productivity per hour worked in CEE countries (US=100)

period of catch-up by the NMS was interrupted by the GFC, but it could be argued that this break in the process of long-term technology-based catch-up is only temporary.

What is important is whether the 1999–2008 spurt has been accompanied by technology catch-up in relation to the world leader. Real technology catch-up should be accompanied by numerous signals of technology accumulation and upgrading: we can detect some of these signals, but based largely on macro or aggregate data.

Figure 2.3 suggests that labour productivity growth was a region-specific rather than a country-specific phenomenon. Differences among CEE countries in increased rate of catch-up to the US are small and follow a common regional trend. Table 2.1 shows that labour productivity for the EU10 narrowed from 36.9 per cent (1997) to 51.4 per cent (2008) of the EU15. Differences among the CEE countries are relatively small, with standard deviations of rates remaining largely the same throughout this period (see Table 2.1). This would suggest that, although there are important inter-country differences in the CEE, they are less significant for catch-up. Historical experience suggests that catch-up is more often country-specific than region-specific, which is an indication that the productivity catch-up experienced by the CEE NMS in the early 2000s was a growth spurt rather than sustainable technology-based catch-up.

The narrowing of the productivity gap was enabled by imports of capital equipment and foreign direct investment (FDI), and a high share

Table 2.1   Labour productivity per hour worked: GDP in purchasing power standards (EU15 = 100)

	1997	1998	1999	2000	2001	2002	2003	2004	2005	2006	2007	2008	Index 2000/08
Slovenia									72.4	73.6	74.5	73.5	
Slovakia	43.2	45.5	45.9	46.7	49.4	52.5	55	55.4	56.9	59.6	63.4	65.8	1.41
Czech Republic	43.7	43.4	44.2	43.9	47.1	47.2	49.8	51.1	51.1	51.7	53.8	53.9	1.23
Hungary	45.0	46.2	44.9	41.0	45.1	46.8	48.3	49.6	49.6	49.9	50.4	52.7	1.29
Estonia				34.5	35.6	37.4	40.1	42.2	44.1	44.9	47.8	47.6	1.38
Lithuania	32.4	33.5	34.2	33.8	37.5	38.8	42.4	43.2	42.7	44.4	46.2	47.8	1.41
Poland	34.7	35.5		38.8	39.3	41.1	42.0	43.3	43.2	42.8	43.6	44	1.13
Latvia				26.6	27.8	29.0	30.0	31.8	32.7	33.6	35.9	38.3	1.44
Romania			18.5	18.6	20.2	22.9	24.7	27.4	28.4	31.0	33.8	39.3	2.11
Bulgaria	22.5	24.1	25.1	27.1	27.8	29.1	29.7	29.7	29.6	30.6	30.9		1.14
Average	36.9	38.0	35.5	34.6	36.6	38.3	40.2	41.5	45.1	46.2	48.0	51.4	
Standard deviation*				8.6	9.3	9.3	9.7	9.4	9.4	9.4	9.8	8.3	

Notes:
*To ensure consistency of country coverage, we exclude Slovenia.
Date of extraction: 03 May 2010.

Source: Eurostat Structural Indicators.

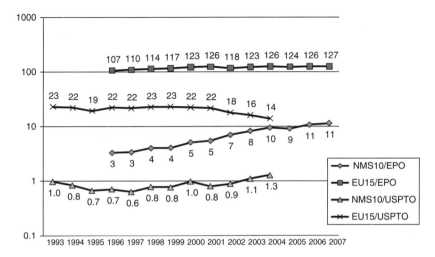

Note: Date of extraction: 24 June 2010.

Source: Based on Eurostat Structural Indicators.

Figure 2.4 *US patents granted and EPO patent applications per million*
EU15 and NMS10 population, relative to the US=100
(logarithmic scale)

of total factor productivity (TFP) or greater allocative and organizational efficiencies. Some of this productivity growth was driven by greatly increased domestic demand, which benefited the financial sector in the form of a credit boom, and an expansion in real estate which affected the construction industry.[3]

The gap with the US, the technology leader, measured as US patenting, is the biggest among world frontier technology efforts (Figure 2.4), and is equivalent to 1 per cent of US per capita levels, indicating that the NMS barely figure as generators of world frontier patentable innovation. The gap with European patenting is much smaller (11 per cent), but still substantial – and especially compared to the productivity gap.[4] This prompts questions about what it is that has promoted the rise in labour productivity if it is not world frontier technological innovation. In line with several other authors (see Kravtsova and Radosevic, 2009; Majcen et al., 2009), we argue here that CEE countries' growth is based on production capability (that is the capability to produce at world productivity levels in a given technology) rather than technology capability.

The lag in EU patenting has been reduced, mainly due to patenting by

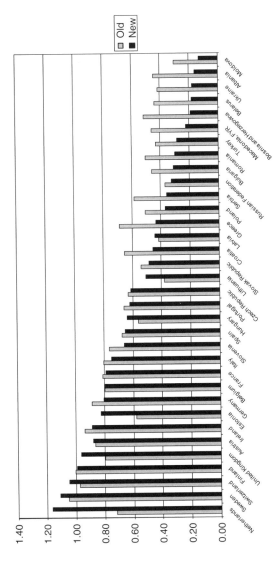

Note: Old technologies: Agricultural machinery, tractors per 100 sq. km of arable land; International voice traffic (minutes per person); Electric power consumption (kWh per capita); Telephone mainlines (per 100 people); Manufactures exports (% of merchandise exports); Air transport, registered carrier departures worldwide. New technologies: Internet users (per 100 people); Personal computers (per 100 people); Mobile phone subscribers (per 100 people); Broadband subscribers (per 100 people); High-technology exports (% of manufactured exports).

Source: Based on World Bank Development Indicators database.

Figure 2.5 Technology gap between European 'core' and 'periphery' for diffusion of new and old technologies compared to the US=1, 2003–05

multinational corporations (MNC) located in the CEE countries (Goldberg et al., 2008). This has led to a change from 3 per cent to 11 per cent of EU per capita levels. At the same time, per capita patenting in the old EU15 in relation to the US has declined. This is a bad sign because it suggests that future technology pull, which potentially could be based on technological cooperation between the East and West EU countries, may not achieve sufficient momentum. A gap in labour productivity of around 40 per cent of the US level confirms that the sources of growth, inevitable in the case of catching-up economies, are related to imitation and the import of capital equipment, rather than to world frontier technology efforts.

It is the diffusion of new technologies, not world frontier technology effort, that is essential for catching up. From a Schumpeterian perspective, the wider diffusion of new technologies would provide greater opportunities for growth than the diffusion of old technologies, which have less growth and productivity potential. Figure 2.5 compares the diffusion of old and new technologies in relation to the US as the technology leader, based on a composite average of several typical old and new technologies. The CEE countries, except Slovenia, are clearly a low class group based on the diffusion of old technologies. For the diffusion of new technologies, there are differences. Estonia and Slovenia are ahead of the other CEE countries, while levels in the CEE as a whole are similar to those in Greece, Portugal and Spain. If we take a cut-off point of 50 per cent, then the CEE countries are doing better for diffusion of new than of old technologies. Levels of technology diffusion in CEE differ more for new than for old technologies, with quite substantial differences among countries, ranging from 82 per cent of the US level in Estonia, to 30 per cent in Romania. The diffusion of new technologies in the 'cohesion' EU economies (Greece, Portugal and Spain) is sufficiently similar to that in the CEE countries such that CEE does not comprise a separate bloc. The wide differences among CEE countries for diffusion of new technologies are indicative of differences in absorptive capacity, which is essential for catching up.

Since they are catching-up economies, CEE countries' growth is based mainly on imported knowledge and technology. Knell (2008) uses input–output data to calculate the R&D embodied in imported inputs and capital goods. His analysis shows that CEE countries' imported inputs and imported capital goods have very high shares of embodied R&D. Reinstaller and Unterlass (Chapter 4 in this volume) show a high indirect R&D content predominating in the NMS. The important role of indirect R&D (that is R&D embodied in imported equipment) compared to direct R&D, has important implications for policy. It suggests that there should be an emphasis on technology transfer policy in countries that rely on technology upgrading via imported technologies. However,

18 Challenges for European innovation policy

Table 2.2   Turnover from innovation as percentage of total turnover

	2004	2006
EU 10 New	12.5	12.4
EU15 Old	12.5	13.5

Note: Date of extraction: 8 June 2010.

Source: Eurostat.

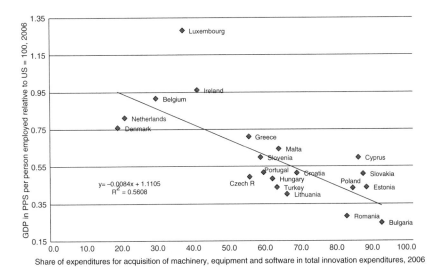

Note: Date of extraction: 20 June 2010.

Source: Based on Eurostat.

Figure 2.6   Relationship between embodied investments and labour productivity in European countries, 2006

this is not to say that these countries do not or should not innovate: catching up assumes high rates of innovation especially in products and processes new to the firm and to the market, although not necessarily to the world. EU innovation surveys show that the share of turnover generated by innovation activities in the old and the new EU member states is very similar on average (Table 2.2). However, the structure of innovation expenditure in the old and new EU is very different and reflects the different nature of these countries' technology efforts. Among the main categories of innovation expenditure are investment in machinery and equipment (embodied knowledge), and investment in R&D (intangible knowledge).

Figure 2.6 shows that the share of investment in machinery and equipment (including software), proxied by labour productivity, is closely aligned to the distance of countries from the technology frontier. Shares of expenditure on machinery, equipment and software in the CEE countries range from 55 per cent (Czech Republic) to 91 per cent (Bulgaria). The share of expenditure on R&D and other intangibles is significantly higher in the more developed EU countries. The lower the level of development of the country, the more likely it is that innovation expenditure will focus on innovation based on imported equipment and its successful operation, and a limited intangible (R&D) component.

Also, the more that innovation in enterprises is focused around equipment, the less will be the focus of these firms on protecting knowledge, which will be in the form of know-how rather than intellectual property. Know-how is related largely to the efficient operation and adaptation of the newly introduced equipment rather than to R&D-based innovation, which results in less importance assigned to patent protection, registered trademarks and industrial design copyrights. In part, this is confirmed by the data in Table 2.3, which show that innovation in NMS is mostly not protected. Differences among countries are probably related to levels of productivity, the position on the innovation–imitation continuum, and differences in industrial and economic structures. Again, this points to the specific nature of the technology efforts in the NMS, which are behind the technology frontier, compared to the efforts in EU economies which are closer to the technology frontier.

The above micro-evidence shows that CEE economies operate behind the technology frontier and do not participate in world frontier technology activities to any significant extent. The frequency and commercial importance of their innovation is similar to the developed EU, but their innovative activity focuses mainly on improvements and adaptations to purchased equipment, which has large indirect R&D content, but promotes limited direct R&D activity.

This picture of Europe as a technologically diverse region is confirmed by several large surveys that evaluate differences in national drivers of growth. For example, data from the World Economic Forum (WEF) for Global Competitiveness Report 2009 show that the EU27 is composed of three groups of countries: those with growth based on achieving efficiencies in a given technology (Bulgaria and Romania); those in transit to growth driven by innovation (6 out of 10 NMS); and those whose growth is innovation-driven (Slovenia, Estonia and the old EU15). Growth in

Table 2.3 Protection methods used by enterprises as a percentage of innovative enterprises by country, EU27 member states and Norway*

	Protection	Not protection
Hungary	**22.7**	**77.3**
Belgium	32.2	67.8
Portugal	33.7	66.3
Romania	**34.8**	**65.2**
Slovakia	**35.2**	**64.8**
Bulgaria	**36.8**	**63.1**
Czech Republic	**38.1**	**61.9**
Estonia	**38.5**	**61.5**
Italy	38.6	61.4
Poland	**40.2**	**59.8**
Greece	**42.3**	**57.7**
Netherlands	42.5	57.5
Lithuania	**44.5**	**55.5**
Spain	45.2	54.8
Finland	50.0	50.0
Luxembourg	51.5	48.5
Ireland	52.0	48.0
Denmark	63.9	36.1
Germany	65.2	34.8
France	83.8	16.2

Note: *Protection refers to patent applications, trademark registration, design registrations and copyrights.

Source: Based on Eurostat, and Bernard (2007).

the EU27 is not based on the availability of factors such as land, natural resources or low-cost labour. It should be borne in mind that the WEF methodology for classifying countries takes account of a variety of both technological and institutional factors. Since institutions are endogenous variables they may be the outcomes of earlier developments, rather than the causes of current development.

In order to measure progress towards innovation-based growth, the EC developed the European Innovation Scoreboard (EIS) Summary Innovation Index (SII), which is a composite indicator that shows the extent to which growth in a country is based on innovation (EC, 2009b). The SII is not necessarily related to national economic growth, especially in the short term, but shows the degree to which a country's economic growth embodies innovation. Growth can be based on other factors, such as economic efficiency, which is reflected not in innovation, but in production capability. Growth may also be based on low-cost or abundant labour or natural resources. The SII score shows whether NMS are 'moderate innovators' or 'catching up' countries (see Chapter 5 by Reid in this volume). This ranking is compatible with the WEF classifications which show that eight out of ten NMS are growing based on efficiencies, or are in transit to innovation-based growth.

Despite some methodological and conceptual differences, the evidence shows that NMS economies are behind the technology frontier and that their growth is based on non-R&D sources of productivity improvements and embodied knowledge diffusion as opposed to knowledge generation. A majority of the CEE countries are still growing based on production as opposed to technological capability. By this we mean that they compete based on efficient use of standard technologies, that is, production capability and adoption of foreign technologies. Econometric research, based on the determinants of the productivity of FDI subsidiaries in 420 firms in five central European countries, shows that a significant portion of their productivity is explained by production capability (Majcen et al., 2009; Kravtsova and Radosevic, 2009). This is compatible with the results reported in Chapter 4 in this volume, which show that there is a weak link between labour productivity and product innovation in the NMS. There is also some evidence (McGowan et al., 2004; Radosevic and Sachwald, 2005) that the sources of productivity improvements in CEE are global value chains and production capability, not technological capability. These results are in line with the broader picture on industrial upgrading in catching-up countries. Hausman et al. (2005) show that countries converge with the international leaders in production quality (measured by unit prices) at an annual rate of 5–6 per cent unconditionally. So, achievement of production capability, that is, upgrading quality in existing products, seems to be a kind of 'automatic' process. At the micro level or the level of company managers and employees, this automatism might seem far-fetched. In reality, it is the outcome of firm-level, active production capability learning, which, in the case of NMS, is also 'FDI assisted' (see Narula, Chapter 6 in this volume). Within FDI assisted we include indirect or non-equity linkages via subcontracting-driven mastery of production capability.[5]

The various evidence points to CEE as a group of countries behind the technology frontier, where technology transfer is very important for growth. If these countries are to grow further and close the gap with the developed EU countries, they will have to invest in their own R&D as well as exploiting imported technology. These differences in the nature and role of technology activities have major implications for innovation policy.

In what follows we point to some alternative theoretical and conceptual thinking that could inform future policies.

2.3   NEO-SCHUMPETERIAN PERSPECTIVE ON INNOVATION POLICY: RATIONALE AND APPROACH

A common feature of Schumpeterian or neo-Schumpeterian approaches is that policy challenges are inextricably linked to the nature of technology, that is, to two key dimensions of technology that are important for growth: distance between the technology leaders and technology followers, and stage in the technology life cycle. In the introduction to this chapter we described some of the key features of the neo-Schumpeterian approach, which assumes that countries that differ in terms of distance to the technology frontier should adopt different policies.[6] Here, we incorporate a complementary perspective, which, in the true neo-Schumpeterian tradition, argues that structural socioeconomic changes are closely linked to technology changes in terms of technology generation and its subsequent diffusion. This perspective is depicted very well in work by Perez (1983, 2006) and Freeman and Perez (1988). Unlike the perspective in Aghion et al., which is based on formalized growth theory, Perez's work, methodologically, is in the historical-structuralist tradition.

The Perezian view of growth departs from the notion of a techno-economic paradigm within which growth is explained as the co-evolution of technologies and institutions (Perez, 1983, 2007). The processes of technology and innovation have common structural features (regularities) and their interactions within the institutional framework are important for understanding the prospects for and obstacles to long-term growth. According to Perez, we are currently in a period that marks the initial phase of a fifth long wave period (the information and communication technology (ICT) paradigm), which is characterized by paradox or a temporary mismatch between the techno-economic and the socio-institutional systems (see also Castellacci, 2006).[7] This framework powerfully underlines a number of organizational and strategic challenges linked to the diffusion of the ICT paradigm. In this perspective, policy issues are explicit, and have been developed in several papers by Carlota Perez (see Perez, 2006, 2001, 2000), and the driving force of full deployment of the potential from the new ICT and its applications, is institutional innovation. Three areas of policy innovation in particular are important: regulation, especially of finance because its role changes in the course of technology cycles or revolutions; re-specialization in developed and developing countries by regions, as new technologies create new technological opportunities; national and global social policies as a way to generate demand for the absorption of new technologies in the deployment stages of a technology cycle (Perez, 2006). As structural change evolves from installation to deployment, this requires that policy accommodates it through a focus on a moving target. This means that policy will be shaped by the nature of the continually changing techno-economic paradigm.

Aghion and Howitt (2005), on the other hand, focus on the economic implications of a country's distance from the technology frontier. For example, Acemoglu et al. (2002: 41) argue that 'there are certain marked differences in the economic organization of technological leaders and technological followers . . . while technological leaders follow an innovation-based strategy, technological followers adopt an investment-based strategy of growth'. The main implication of this view is that national growth policies need to vary according to the distance from the technology frontier and depending on the institutions (Aghion and Howitt, 2005). Similar arguments were proposed by the so-called Stiglitz group (Cimoli et al., 2007: 20), which argues that the stringency of the new international constraints 'is likely to vary from sector to sector and from technology to technology *and it is likely to depend also on the distance of any country from the international technological frontier'* (emphasis added). Aghion and Howitt (2005: 42) develop this line of reasoning by arguing that *'growth-maximizing policies* (e.g. competition and entry policies, the allocation of education funding or the design of macroeconomic policies) *should vary with a country's or sector's distance to the technological frontier, and/or with the country's level of financial development'* (emphases added).

Aghion et al.'s and Perez's perspectives are complementary since, basically, they address two different but interconnected effects of technological change on growth. Aghion et al.'s concern with technology distance and its policy implications is relevant for achieving a better understanding of the policy issues in countries that are considered to be on the technological 'periphery' as opposed to those considered to be part of the technological 'core'. In the European context, and in terms of technology generation, developed EU countries de facto comprise the technological core and the CEE countries represent the periphery. Complementing this, Perez argues that policies should differ depending on whether the techno-economic paradigm is in its installation or deployment stage. Although an oversimplification, it could be argued that in the installation stage, the policy focus should be on the supply side, on innovation and structural change, and that in the deployment stage, promoting diffusion and demand for technology uptake and absorption should be the major policy concern. Table 2.4 summarizes these complementarities.

Table 2.4   (Neo)Schumpeterian perspectives on technology and growth and their policy implications

| Aghion et al. (distance from technology frontier) | Perez (stages in transformation of techno-economic paradigm) | |
	Installation period	Deployment period
Core	− Supply-oriented policies (technology push) − Policy oriented towards new sectors with big technological opportunities ('new economy') − Structural change in policy focus − Innovation-based strategy	− Demand-oriented policies − Innovative use of new technologies to create social welfare
Periphery	− Supply-oriented policies adapted to periphery − Absorptive capacity for new technologies − Search for narrow 'windows of opportunity' to catch up with innovation leaders − Investment-based strategy	− Diffusion-oriented policies − Adaptation and effective use of new technologies − Investment-based strategy

In summary, both perspectives consider that the policy challenges are inextricably linked to changes in the type of knowledge countries need to accumulate and improve. For Aghion et al. the type of knowledge differs depending on the distance to the technology frontier, while for Perez it differs depending on the stage of the technology cycle. As is the case with any high-level stylizations, these two perspectives are not directly 'convertible' to policy. Nevertheless, they serve as useful heuristics and throw new light on the policy issues. They provide several points of departure for thinking about innovation policy.

First, the EU27 is a country grouping encompassing different technological levels or different distances from the technology frontier. This diversity of 'production functions' requires new policy approaches and possibilities to exploit potential synergies and diversities. The national technology leaders and those behind the technology frontier interact in both upstream (R&D, technology generation) and downstream (production, services, technology use) value chain activities. These links are forged through a variety of commercial, capital and knowledge/personal channels. For example, R&D groups in CEE are already integrated in EU networks of excellence, while FDI from the developed EU has already transformed certain industries in CEE such as automotives and electronics. The issue for policy is how to enhance these interactions so that they lead to complementarities and synergies in knowledge genera- tion and diffusion in both the EU core and the periphery.[8] From a (neo) Schumpeterian perspective, innovation policy convergence is not desir- able in conditions of diverse production functions which require that policy should be tailored to different modes of technology accumulation (innovation and imitation) in different countries and sectors. Growth policies should be country-specific in their dependence on the distance to the technology frontier (innovators/imitators) (cf. Aghion et al.). The focus of innovation policy for the technology leaders needs to be dif- ferent from the policies applied to technology followers or catching-up economies.

Second, innovation and imitation are endogenously determined by the national system of innovation (NSI) (including the financial system) and competition, but should depend also on the distance to the technol- ogy frontier. Whether a country is a leading innovator or imitator will affect its optimal markets and institutional structures. The institutions appropriate for the accumulation and adaptation of foreign generated technological knowledge, will differ from those required for the generation of technology frontier-level knowledge. For example, in countries behind the technology frontier, where supply of and demand for innovation is low, policy should be addressed to market forces and focus on framework- supporting regulation, rather than being interventionist. At the same time, weak domestic firms require policies that will support their growth and that of the related infrastructure. National technology leaders should focus on explicit innovation policy designed to achieve or maintain their technological positions in the face of technological uncertainties and high R&D upfront costs. On the other hand, the framework conditions for markets and goods are essential *filtering mechanisms* for the selection of new technologies. In practice, a balance between these two options rather than an exclusive focus on one or the other, will be optimal (see Chapter 5 in this volume). This balance will depend on the distance from the tech- nology frontier and on the stage of the technology cycle, or phase in the techno-economic paradigm (see below). However, in both cases the scope of innovation policy should be broader than the conventional perception. Frequently, the major challenges in the development of innovations are generated by factors that are beyond the scope of an explicit innovation policy.

Third, policy should take account of the stage of the techno-economic paradigm or technology cycle. A focus on innovation and structural change that concentrates on specific sectoral patterns of structural change may not be the most appropriate for the technology cycle deployment stages when issues related to the diffusion and absorption of technology in peripheral sectors are important. However, a focus on structural change is important in the initial or installation stages of a new techno-economic paradigm. In these stages, there are potential windows of opportunity and latecomer advantages (Veblen, 1915) based on newcomers not being constrained by the limitations of old technologies (Perez and Soete, 1988). It is important to understand when the industry structure is becoming a constraint on growth and when structural weaknesses can be compensated for by increased technology intensity (for a further discussion see Chapter 4 in this volume). This is very relevant to the debate on the EU–US productivity gap. It is important to understand whether EU productivity is lagging due to greater specialization in medium-rather than high-tech industries compared to the US, or whether the EU lacks technology dynamism due to an 'inferior' innovation system (see EC, 2004). In a Perezian perspective this should be seen in the context of the changing techno-economic paradigm. Structural issues may be relevant in the installation stages of the current techno-economic paradigm (1980s–1990s–2010s), but not in its deployment stages (2000s–2010s–2020s). Hence, policy that does not capture the changing nature of the techno-economic paradigm is chasing yesterday's targets.

What follows from all three propositions is that innovation policies appropriate for the EU27 should be country- or country group-specific, not convergent. This differentiation is dependent on the technological position of the country or region being considered (technology core or periphery), and also on the stage (installation or deployment) in the technology cycle of the country or industry. Overall, the (neo)Schumpeterian approach argues that technology diversity demands policy diversity. We must avoid policy myopia or an excessive focus on an EU policy mix that addresses issues of relevance to the technology leaders, but which is not appropriate for moderate innovators or catching-up countries (Nauwelaers, 2009).

We develop these propositions through an examination of several policy related issues in the EU27 context, most of which are discussed further in the contributions to this volume.

2.4   POLICY ISSUES ARISING FROM A (NEO) SCHUMPETERIAN APPROACH

We can identify six key propositions related to country- or country group-specific policy:

- different patterns of technology upgrading in CEE countries matter for their innovation policy;
- different system failures in the CEE EU demand different policy priorities;
- policy needs to balance the trade-off between support for excellence versus local relevance;
- FDI policy should be an integral part of national innovation policy;
- establishing missing demand-oriented policies;
- framework conditions, especially competition, matter for innovation in relation to 'sectoral policy regimes'.

2.4.1 Different Patterns and Policy Implications of Technology Upgrading in CEE

The CEE countries are predominantly technology users, based on high levels of indirect technology, that is, high share of knowledge embodied in imported equipment (see Figure 2.6, and Chapter 4 in this volume; Knell, 2008). They have low own-R&D intensity and rely on R&D embodied in imported inputs. Technology transfer and its effective absorption are probably the most important routes to technology upgrading. However, the dominant policy metrics represented by the EIS is based on world frontier activities and the accompanying R&D and institutional requirements. Technology upgrading, according to the EIS, evolves as a result of increasing investment in R&D, high-level technology, and patents as its outcomes, in the shift from catch-up country, to moderate innovator, and then to follower and leading innovator. However, the pattern of technology upgrading of economies behind the technology frontier is different in practice. Based on similar data to Figure 4.1 in Chapter 4 in this volume, the pattern of technology upgrading in CEE countries evolves along an alternative or non-EIS pattern, as described in Box 2.1.

This pattern of technology upgrading requires a change in the policy focus from world excellence and world technology frontier efforts, implicit in the EIS metrics. These changes include a focus on: a) the efficiency and effectiveness of technology imports embodied in imported equipment or subcontracting relationships; and b) interactions among and coupling of domestic technology/R&D efforts and imported embodied and disembodied knowledge.

An examination of the current innovation policy mix in CEE countries suggests that, at least in some countries (cf. Slovakia), policy is focused relatively more on technology transfer and human capital through the contributions of the SF (Balaz, 2008). Therefore, in these countries, it would seem that the focus should be on equipment purchase, but linked

---

BOX 2.1   NON-EIS PATTERN OF TECHNOLOGY UPGRADING

1st stage: *low overall technology intensity.* In this stage both direct and indirect R&D and other innovative activities are at comparatively low levels.

2nd stage: *high indirect technology intensity.* In this stage R&D intensity is low, but R&D embodied in imported equipment increases.

3rd stage: *average direct and indirect technology intensity.* Further upgrading requires the coupling of imported knowledge with own R&D activities, which increase to an average level.

4th stage: *high direct technology intensity.* As R&D intensity increases, the relative share of indirect technology effort decreases and the country reaches the world technology frontier.

---

to manufacturing advisory services, changes to the work-place organization, training, and so on. Simply providing grants for companies to upgrade is not enough.[9] However, in the majority of the CEE countries, the policy priorities are human resources for R&D, R&D–industry cooperation, and business sector applied R&D (see Chapter 5 in this volume). These policies are related not to technology imports, but to R&D focused growth. They reflect the EIS policy framework, which is focused around the R&D and innovation infrastructure for new technology-based firms and shows that policy makers have identified the challenges based on the dominant (EIS) metrics. What is important, however, is whether this is the right metrics. Evidence on the technology position of CEE NMS suggests that it is not. The implications for policy of a non-EIS pattern of technology upgrading are somewhat different. In order to advance from exclusively imported technology-based productivity growth, these countries will need to complement purchases of embodied technology with 'soft' investments – in training and design (Chapter 5 in this volume). There also needs to be a focus on the knowledge links between FDI and local firms, and between large and small and medium enterprises (SME) (Chapter 6 in this volume), and on enhancing and stimulating demand for local R&D (Chapter 7 in this volume). Human resource policies should concentrate on lifelong learning and enterprise- and technology-specific skills rather than provision of human resources for the R&D sector.

2.4.2   Different Types of Failures in CEE NMS Require Different Policy Priorities

Advances in innovation policy analysis have been achieved in terms of understanding types of failures in innovation promotion. There has been a shift from a narrow focus on market failure, to a focus on the diversity of failures in both markets and the institutional set-ups that form these markets and the institutions that underpin them, and in the capabilities of firms and the systems in which they operate (Arnold, 2004; Negro and Hekkert, 2010). Reid (Chapter 5) provides an analysis of these issues and shows that CEE countries' policies tend to focus relatively more on direct financial intervention in the applied R&D, research and innovation infrastructures. In other words, they focus more on 'capability' and 'institutional failures' relative to the high share of policy instruments that address 'framework failures'. Combined with an 'investment driven' approach to research, technology, development and innovation policy via technology acquisition by SME, this is a good reflection of the position of these countries in relation to the technology frontier.

Chapter 5 shows that there are similarities across the CEE5 (Czech Republic, Hungary, Poland, Slovakia, Slovenia) in terms of the focus on technology transfer and hard technology, but that there are significant differences across countries in terms of policy profiles. A country-specific approach, such as the one proposed by the different contributions in this volume, shows that there are: a) commonalities in the policy focuses of the CEE countries based on their similar distances from the technology frontier; and b) different institutional structural conditions in each country which require different policy instruments. A country differentiated approach must be a mix of commonalities in terms of 'distance to the technology frontier' and 'institutional differences' based on the different national political economies.

SF are a common feature of the CEE countries. They are key mechanisms facilitating interaction with the EC. At the time of writing, they were being underexploited in efforts to promote a shift towards innovation-driven growth (see Chapter 5). One of the constraints to their greater effectiveness was and is a lack of endogenous capacity to design and implement research, technology and development (RTD) projects (Barca, 2009).

There is a need for better coordination between a cohesion policy (delivered via the SF) and the EC Competitiveness and Innovation Programme and its research Framework Programmes (FP). However, it is unclear whether better connection between the SF and the FP would lead to anything more than new RTD capacity, not necessarily relevant to the local economy. In other words, better coordination may not by itself resolve locally relevant issues unless the approach to exploitation of SF is re-examined. There is evidence that the incentive provided by the SF stimulates financial absorption more than it ensures long-term effects. The accompanying bureaucracy is often further complicated by the rules of national administrations, and this reduces the effects of the funds on increasing national innovation capacity (Technopolis, 2006). There is a need for a re-evaluation of the role of SF in building firm- and industry-specific infrastructures with strong linkages to public research systems as a means of attracting additional or embedding existing FDI. This re-evaluation should be approached from the perspective of growth rather than redistribution.

SF are used to substitute for national funding of RTD in the CEE NMS. There are limits to this substitutive role, and re-conversion into complementary funding would be a better way of ensuring local relevance. The EC FP should make a major contribution to R&D quality in NMS, but increased FP participation will not necessarily lead to increased funding efforts at national level. In both cases, achieving complementary national funding will be easier if overt efforts are being made to ensure local relevance. This may lead to smaller financial benefits for NMS, but higher local relevance of funded RTD activities.

2.4.3 How to Achieve a Balance between R&D Excellence and Relevance

Greater integration of local NMS R&D groups into EU R&D networks should lead to improved quality and even international excellence of R&D systems, but not necessarily greater local relevance (Radosevic and Lepori, 2009). It may widen the gap between supply of and demand for local R&D and lead to a situation where there are islands of international excellence that are unrelated to areas of local demand for R&D.

We can expect that the 'top performers' in NMS R&D systems will be strongly integrated into the European Research Area. This should have positive effects in terms of R&D dynamism and excellence since it will mean that the best R&D groups will be 'plugged' into EU R&D networks. In the long term this should have positive effects on R&D in the NMS through the application of the criterion of international excellence. However, none of this will ensure that these countries' R&D systems will be relevant to their local economies. As the best R&D groups become integrated into EU networks, the gaps with the local business sectors may widen. The situation among CEE NMS may come to resemble the situation in Greece where there is a competent R&D system with relatively few links to the domestic business sector (Collins and Pontikakis, 2006). It is essential that the EU and the NMS should not replicate this scenario.

A common feature of most NMS is that improvements in their NIS are related largely to research and are reflected in publishing activity (OST, 2004). This trend will be reinforced by the Europeanization of their R&D systems. We can expect improvements in terms of a better balance between incentives (selection through project funding) and stability (share of institutional funding). These aspects have been very unbalanced, which has resulted in seemingly hyper-competitive R&D systems that do not contribute to stability, or overly stable systems with no incentives for the achievement of excellence. The major bottleneck, constituted of weak domestic demand for R&D coupled with a weak business enterprise sector, is likely to remain a major structural weakness of these countries' R&D systems.

The R&D systems in NMS are undergoing strong Europeanization, which means that the dynamics of EU RTD policy is becoming part of the organizational logic of national science, technology and innovation policies. This should bring benefits in terms of fostering international excellence and institutional convergence in R&D evaluation. However, a strong impact on the definition of policy priorities and what can be considered 'legitimate' issues may lead to the mechanical transfer of policy models, which may not be relevant (Radosevic, 2004a). The experience of Europeanization in the southern EU countries shows that the strongest effects have resulted from policy actions and mechanisms directed to national priorities (Featherstone and Kazamias, 2001). There is no reason to believe that the case of the CEE NMS will be any different. The outcome could be strong policy myopia or a situation where the importance of local R&D issues and the search for local solutions is ignored. An excessive focus on established EU policy at the expense of local issues may reduce the pressures to develop policy within a process of discovery, that is, through the search for local solutions.

Improvements in the CEE NMS overall have been achieved through the introduction of peer review-based funding mechanisms geared to the achievement of scientific excellence. However, an exclusive focus on the criterion of excellence could lock scientific specialization into areas of past excellence (Kozlowski et al., 1999). Funding systems are inadequate to ensure industry and social relevance, which require much stronger involvement of users in evaluation and funding (EC, 2009c).

2.4.4 Integrating FDI and Innovation Policy

Because they are behind the technology frontier CEE countries have huge opportunities to upgrade technologically through cooperation with foreign companies and import of embodied technology. However, effective acquisition of foreign technology requires absorptive capacity. Foreign technology cannot substitute for poor domestic absorptive capacity, but can complement domestic technology upgrading (Radosevic, 1999a). CEE countries have weak domestic firm-level capabilities inherited from the socialist period; these capabilities were further eroded as local firms experienced break-up, privatization or prolonged deterioration (Radosevic, 1999b). Firms that during socialism were mostly production rather than business units were subjected to foreign takeovers, which reduced them to single activity affiliates – probably reflecting the market value of the inherited technical and organizational capabilities and the strategies of the MNC (Radosevic, 2004a). There were expectations that FDI would generate substantial direct and spillover effects. The former have materialized through the increased contribution of FDI to employment and value added (Jindra, 2005), but spillover effects have been much smaller than was hoped. Upstream spillovers (interactions with suppliers) have been positive, but downstream spillovers (links to buyers) have been negative, and horizontal spillovers (interactions with competitors) insignificant (for an overview, see Jindra, 2005).

Some recent research points to the essential role of firms' absorptive capabilities in generating knowledge spillovers from MNE (Damijan et al., 2008). This confirms the view that local firms' weak absorptive capabilities are a key aspect of weak spillovers (technology transfer) from MNE. In CEE countries, mutual FDI linkages may be the only way to achieve technology upgrading. From a policy perspective, FDI policy and its integration with innovation policy needs more emphasis. As argued by Narula (Chapter 6 in this volume) MNE, industry and innovation policies need to be inextricably linked. The challenge is to embed MNE into the local economy and use global value chains as opportunities and mechanisms for technological upgrading. Thus far, the focus too often has been on capital flows and employment at the expense of linkages (McGowan et al., 2004). How to enhance the domestic knowledge gained from adapting and customizing imported capital goods is a central challenge for innovation policy. It requires policy innovation and experimentation since there are no ready-made EU policy solutions.

The NMS should try to integrate FDI and innovation policy – organizationally, financially and in policy terms. The overall objective should be to orient the SF to build industry-specific and technology-specific support services. This will require foreign investment agencies being pushed to focus on embedding foreign investors, and further developing FDI rather than just attracting new investment. Embedding investment should be seen as a continual activity aimed at integrating FDI into local economies to stimulate multiple linkages by opening up opportunities for industrial upgrading for both domestic and foreign firms. It would be useful to explore the rationale for and feasibility of establishing a coordinating and facilitating body to support trans-European value chains and supply networks, and link them to global value chains. We are aware that, from a political viewpoint, this may not be seen as important, but from the perspective of countries at the EU 'periphery' it is a highly relevant issue.

McGowan et al. (2004) developed a somewhat heretical proposal that the EC should hold public (open) contests for FDI to provide incentives for private and public actors to develop innovative solutions together, to improve the investment climate at sub-national level, and to provide matching grants at EU level. It was envisaged that these contests could serve as:

- incentive devices for local government and domestic firms to engage in meaningful joint action and reform;
- coordination devices to coordinate activities at national and EU level under the umbrella of private–public competitiveness projects;
- mechanisms to share policy knowledge.

The idea was to provide a platform to integrate innovation policy objectives into FDI policy, and to coordinate this policy across EU, national and regional levels.

2.4.5 Demand-oriented Innovation Policy

Demand for technology is not the same as market demand. Demand for technology operates through an organizational structure and, hence, the organizational features of the economy, such as firm size, firm interactions and the institutional context, play important roles. Innovation is an uncertain process in which the user's needs are not clearly articulated or are unknown. Hence, an institutional context that hinders user–producer interaction makes it difficult to understand the demand for technology. It is for this reason that the concept of demand for technology is undeveloped in innovation theory. Schumpeter assumed that demand would automatically adjust to new supply opportunities (Andersen, 2009). This contrasts with the neo-Schumpeterian systems of innovation view, which sees user–supplier interactions as important drivers of the innovation process (von Hippel, 1986; Lundvall, 1992). However, the notion of demand is broader and includes potential demand and unknown users. The role of customers often goes beyond the decision of whether or not to buy (Bhidé, 2008). Consumers do more than just acquire knowledge about a 'standard technique' and often play adventurous or 'entrepreneurial' roles in the design of new products (Bhidé, 2008).

The institutional set-up of the economy or the NSI is important for generating technology, articulating demand and facilitating market activities. It is necessary to be able to recognize who are the (potential) users and how demand for technology is articulated across a range of public and private actors (Kaiser and Kripp, 2010). As technology encompasses public and private elements these must be coordinated to achieve new technologies. There can be coordination problems including information deficiencies, different planning horizons and (mis)matched capabilities, affecting both the quantity and quality of demand. For example, supply and demand for new telecommunication services requires much tighter competition policy that takes account of the innovation and market structures (Edquist and Hommen, 1999). Insufficient demand for standardized services requires a different approach to increasing demand for new, untested telecommunication services for example, and the demand side of innovation policy is comparatively less developed than the supply side (see Chapter 7 in this volume).

Growth and recovery in the CEE countries was not linked during the 1990s to demand for domestic R&D and technology efforts (Radosevic and Auriol, 1999; Radosevic, 2004c). It has been shown that the sources of growth in CEE countries are improved production capability based on imported equipment more than technological or world frontier innovation capabilities (Kravtsova and Radosevic, 2009). It would be misleading to assume that growth could continue in the absence of innovation capability, or that demand for local innovation would be automatic. Research on new knowledge-intensive firms in CEE, for example, indicates that a limited domestic market is the key obstacle to enterprise growth (Radosevic et al., 2010). Pockets of demand for sophisticated local products and local innovation are confined to the government sector and in the main to ICT. The average CEE consumer is not a demanding buyer, and CEE countries' businesses believe that the level of supply of RTD is higher than the level of demand (Radosevic, 2009b). Policies to stimulate demand for locally generated innovation are required, which in turn would facilitate the technological upgrading of local firms. Demand-oriented policies are rare in the EU CEE NMS (see Chapter 7 in this volume).

We have no comprehensive and theoretically rooted explanation for the very weak demand for technology in the EU CEE NMS, but it would seem that structural, institutional and policy factors play an important role. First, in terms of purchasing power, these economies have mostly small and relatively poor national markets, whose demand for technology is quite weak.[10] Second, technology generation and diffusion rely on public–private interactions which, in the case of these economies, are poorly developed or are undergoing reconstruction. Third, policies that would directly stimulate demand for technology, such as regulation and technology public procurement, are either non-existent or are inconsistent with EU rules, which generally do not allow the building of national champions through preferential access to public contracts. We discuss these policy issues below.

Climate change and demand for sustainable development are generally increasing regulation and incentives for environmental innovation. However, for countries behind the technology frontier, regulation and standards cannot be used to promote innovation. Regulation through national standards constitutes an innovation policy instrument primarily for the technology leaders, and in small markets, exclusive national standards are not effective inducement mechanisms.

Public procurement works to stimulate innovation only in the case of world frontier innovators (Edler and Georghiou, 2007). It is a demand-side incentive since the functions of a product or system are specified by the procurer, thus demand 'pulls' innovation through the articulation of a proto-demand (Edquist, 2002). Public procurement is not used extensively in the EU because it favours national champions, which goes against the logic of the Single Market. Also, pressure from buyers for 'value for money' means that innovation is less important. In the case of NMS, the challenge is to support demand for imitative innovation (innovation new to the country, but not new to the world). An obvious solution is to offer some kind of protection or subsidy to local producers, but this would inhibit the functioning of the Single Market which has competing technology suppliers in other parts of the EU.

Some analysts (see Georghiou, 2006) include systemic policies, such as cluster and value chain policies, in demand-side measures. This area of policy is comparatively more developed in the CEE countries. An OECD (2005b: 211) study on clustering in CEE shows that clustering is 'strongly FDI-driven, with local firms clustering around one or several strategic foreign investors, especially in the automotive sector'. Domestic SME clusters, often in traditional sectors, are slowly emerging in all countries, but some critics of cluster policy in CEE maintain that 'many cluster organizations of local SMEs are only virtual clusters looking for subsidies' (Sass et al., 2009: 27). They point to a lack of interest among MNC in developing horizontal cooperation links because they would increase the bargaining power of local suppliers.

Tax subsidies for R&D to stimulate demand are rare in CEE countries. However, some preliminary experience in Slovenia suggests that they may be important as a stimulus for business sector activity. In the year following the introduction of an R&D tax subsidy in Slovenia, R&D investment grew by an impressive 23 per cent, compared to stagnation in real growth in business enterprise R&D in the previous two years (Bucar, 2009).

Overall, technology demand policies have not been tested in CEE NMS, which, in part, is a reflection of their relatively small application in advanced EU countries (see Chapter 7). It is also a reflection of the greater difficulty experienced by countries behind the technology frontier to formulate demand-oriented innovation policy. It would seem too that, in this policy area, the EU regulatory framework works to constrain rather than promote local innovation.

2.4.6   Macroeconomic Stability, Competition and Innovation Promotion

The development of innovation activities requires stable macroeconomic conditions, otherwise enterprises will focus on short-term gains, and investment in innovation will be small and confined mostly to activities that generate very rapid financial returns. Macroeconomic stability is of paramount importance for the CEE countries for two main reasons. First, price stability facilitates the efficient reallocation of resources and highlights unprofitable products, firms and industries. Second, stability is required to attract foreign capital and operate an open trade regime.

Macroeconomic stability is a complex notion that cannot be encompassed by a single variable: it includes prices, exchange rates, budget/current account deficits, and so on. Its effects on industry operate via the 'macroeconomic regime' or the composition of different variables (prices, interest rates, exchange rates, current account and budget deficits), rather than via individual macroeconomic variables. Macroeconomic stability can be achieved by different portfolios (mixes) of the macroeconomic variables constituting country-specific 'macroeconomic regimes'.

On the other hand, technological change is an exemplary case of intensive change in cost structures, which requires accommodative rather than tight macroeconomic policy. More strict macroeconomic regimes are suited to situations where changes in cost intensity are smaller; extensive cost structure changes require that macroeconomic policy accommodates to this change. Macroeconomic stability on its own is not enough to develop national technology potential. The most efficient macroeconomic regimes are those that manage to combine the need for macroeconomic stability with the need to promote structural, that is, technological change. (For a discussion of this link in the context of CEE see Kutlaca and Radosevic, 2006.) These conflicting objectives are difficult to achieve and require closely coordinated macroeconomic and innovation policies.

The link between technology policy and macroeconomic policy and greater coherence and coordination of the incentives for innovation and macroeconomic stability, is neither well defined nor well understood.[11] This has led to a separation between macroeconomic and sectoral policies, whether related to industry or technology. This separation can produce contradictions among objectives and even adverse outcomes[12] because the conversion of macro-conditions into micro-economic incentives is achieved via the incentives and policies in place at the micro- and sectoral levels, or the 'sectoral policy regime'. For example, the diffusion of renewable technologies is influenced more by the specific 'policy regime', which includes ownership and the role of markets, than by narrowly defined programmes for the promotion and diffusion of new, renewable technologies. Regulatory policies, including specific sectoral regulations, are essential (though not recognized) elements of contemporary national innovation policy.[13] Industry-specific policy and enforcement issues are often the main constraints to productivity growth (Palmade, 2005). In a series of studies on different countries and sectors McKinsey Global Institute[14] shows how sectoral policy regimes affect competition and can represent the biggest barriers to economic growth. In what follows, we focus on competition policy as among the most important micro-policies or ingredients in a 'sectoral policy regime'.

Most empirical research demonstrates a positive correlation between innovation and competition (Gianella and Tompson, 2007). Some studies highlight the complexity of this relationship, showing that the effect of competition depends on the distance of firms from the technology frontier (Aghion et al., 2002). Reducing the barriers to entry for foreign products and firms has a more positive effect on the economic performance of firms and industries that initially are closer to the technological frontier, but can be damaging to firms and industries that initially are far behind the frontier (Aghion and Bessonova, 2006). This polarizing effect of liberalization has important consequences for competition policy, which needs to take account of the technological level of the local industry to assess the effects of competition on performance. In policy terms, this requires coordination between competition and industry policy. In a traditional static perspective, there is a conflict between competition and industry policy objectives. However, if we consider innovation as driving industry dynamics, the trade-offs between competition and industry policy may be lower than is often assumed. In the context of innovation, competition policy should be seen as a mechanism 'to foster economic progress through innovation, which induces a kind of dynamic or *selective* efficiency' (Possas and Borges, 2005: 25). In this case, we need to assess 'the extent to which a market, as a selective environment, induces the *evolution* along any innovative trajectory' (Possas and Borges, 2005: 26, emphasis added).

This problem is nicely defined by Gaffard and Quéré (2006) who investigate whether competition policy aims to optimize market structures or encourage innovative behaviours.

The problem is that non-competitive behaviour based on local monopoly may favour innovation while simultaneously being an obstacle to the achievement of static efficiency. So, there is a trade-off between market imperfection and market failure, which is intrinsic to any innovation process. Policy must aim at achieving market imperfection, but without inducing market failure and bearing in mind that innovation is rarely conducted by single firms and usually requires cooperation agreements, which introduces the risk of collusion (Teece, 1992).

Higher rates of entry and exit (higher firm turnover) are generally more growth-enhancing in countries that are closer to the technological frontier (Aghion and Howitt, 2005). In Chapter 3, Aghion et al. conclude that competition is growth-enhancing for the EU CEE countries, most of which are behind the technology frontier although some are advancing towards innovation-based growth. On the other hand, increased competition within the EU Single Market does not automatically trigger an increase in the accumulation of knowledge and innovative capabilities (cf. Cimoli et al., 2007). EU accession has increased competition in the domestic market and increased foreign inflows of, especially, embodied technology. However, accession by itself has not led to increased R&D activity, which has been promoted largely by the contribution of the EU SF (see Chapter 5).

In summary, framework conditions, and especially macroeconomic stability and competition, matter for innovation policy. Schumpeterian economists would probably agree that innovation policy cannot compensate for missing framework conditions. However, there would be less agreement about what these conditions should consist of and how they affect innovation. It seems that there are no across-the-board solutions, and policies require intimate knowledge of innovation and markets, within a dynamic perspective. The literature on the business climate and its role in economic development is large and often confusing (see Carlin and Seabright, 2007). It seems that there are limits to generalizations, and solutions need to be context- (country and sector) specific. If we want to understand why certain innovation policy instruments work or not we need to take account of the particular framework conditions and how they are affecting innovation in firms within specific 'sectoral policy regimes'.

2.5 CONCLUSIONS

This chapter has outlined the rationale for developing a new innovation policy approach for the enlarged EU that takes account of technological and developmental differences across the countries that make up the EU27. We have analysed why these policies need to differ from EU-wide policy approaches.

First, there is a large body of statistical evidence on the diversity of innovation capacities within the EU27. This diversity, and the differences in the underlying drivers of growth, require different policies and different indicators to gauge and benchmark innovation performance. The notion of best practice in this context lacks substance since the specific technology and institutional context matters more.

Second, the diversity of 'production functions' across the enlarged EU represents a potential advantage, which should be exploited. Technological accumulation is a process affected by a variety of factors of which the Single Market and resulting competition constitute just one. Increased competition should be complemented by endogenous R&D to assimilate technology from FDI and exploit technology spillovers and access to complementary assets. This requires a focus on technology supply chains and complementary assets. Policy should foster the development of linkages and the capabilities of local firms (Dantas et al., 2007). We have argued that this will require close integration of FDI and innovation policy (see also Chapter 6).

Third, a new policy approach should go beyond the confines of traditional R&D and innovation policy. These policies are important, but far from sufficient to enhance technology upgrading in countries that are at or behind the technology frontier. In order to capture country specificity in terms of the position vis-à-vis the technology frontier, innovation policy should be complemented by competition, higher education and labour market and financial market policies. Coherence among these policies is essential for innovation policy to be effective since, on its own, it cannot compensate for weaknesses in framework conditions and policies. So, policy mixes must be country-specific. The need to move beyond conventional innovation policy has been recognized in EU policy analysis for some time. For example, the idea that we should try to capture the innovation effects of non-innovation policies is described as 'the third generation innovation policy' (EU, 2002; OECD, 2005a). However, the chapters in this volume show that the evidence for this is scarce. There is little for policy analysts to rely on in their attempt to capture, for example, the relationship between innovation promotion and macroeconomic stability or competition policy.

Fourth, the mix of innovation and complementary policies must be country specific and, in addition to the distance from the technology frontier, should reflect each country's traditions, aspirations and institutional needs. For example, competition policy should play different roles in different countries, and education policies will need to focus on different types of skills. This calls for country specific 'policy mixes' and complementarities between innovation policy, structural reform and other policies. However, there are some common or regional needs among countries. For example, most countries need support in the form of early-stage finance adapted to local conditions and requirements, a need that has become more acute in the current global economic crisis, which has caused many sources of early stage finance to disappear.

Fifth, the Schumpeterian approach that underpins the work in this volume emphasizes policy divergences, which reflect the technological positions of countries or regions (technology core or periphery) and the stage in the technology cycle (installation or deployment). This introduces into policy analysis not yet explored dimensions of innovation policy (competition, FDI), overshadowed by 'one size fits all' or 'best practice' policies. Investigation of these other policy dimensions is becoming urgent in a period when the EU is faced with a radical rethinking of a wide range of its policies – from monetary to research. There needs to be a new policy mix of EU-wide and national and region-specific policies. We must avoid a focus on policy that addresses issues relevant to the technology leaders, but not the moderate innovators or catching-up countries.

Finally, the development path of CEE must not be based only on domestic considerations; it must encompass EU mechanisms including institutional requirements, assistance, monitoring and coordination. The NMS are members of a large institutional country grouping – the EU – which strongly influences their NSI which goes beyond FDI. Bruszt and McDermott (2008) quite insightfully refer to *foreign direct involvement* or the inclusion of a large diversity of external state and non-state actors in assisting and monitoring domestic institutional change in CEE countries. The 'EU transnational institutional regime' has increased the pressures and incentives for institutional change in the CEE countries. However, Europeanization so far has not led to technology accumulation or produced sustainable growth and catch-up. In other words, the EU as a new transnational institutional regime is only partially successful as a development programme. The recent GFC and the example of Greece indicate the limits to its development role. This calls for an examination of the EU's institutional architecture and a re-examination of the relationship between institutional 'openness' and 'autonomy' in national economic growth. Technology accumulation and innovation policy are important aspects of the renegotiation process. We hope that the insights provided by the contributions in this volume will provoke new thinking about these issues.

NOTES

1. I am grateful to Carlota Perez and Despina Kanellou for stimulating discussions on the ideas that inform this chapter, and to Andreas Reinstaller, Alasdair Reid and Anna Kaderabkova for very useful comments on this chapter. All remaining errors are my responsibility.
2. For example in 2003–08 the average annual rate of GDP in the 10 CEE EU NMS was 5.3 per cent compared to 1.9 per cent in the old EU15 (World Bank, 2010).
3. For a good analysis of this type of growth see the Expert Group Report on Estonia (Varblane, 2008).
4. The only exception here is Slovenia, where the gap with the US is only 53 per cent; if Slovenia is excluded, the gap for the NMS is 7 per cent.
5. However, it seems that some CEE countries (Hungary, Croatia, Lithuania, Romania, Slovenia) have less scope for further quality improvements and need to move towards new products (see EBRD, 2008), which suggests a move from production capability learning towards active technology mastery of products and processes new to market.
6. We use the terms Schumpeterian and neo-Schumpeterian interchangeably because much of the existing work does not differentiate. For example Aghion et al. describe their work, which is neo-Schumpeterian in essence, as Schumpeterian.
7. Various authors adopting this perspective were probably inspired by the work of Freeman and Perez, but do not actually refer to it, for example Helpman (1998) and Eliasson et al. (2004).
8. For an analysis along these lines in the context of south-east Europe see Radosevic (2009b).
9. I am grateful to Alasdair Reid for this point.
10. Locally driven innovation in CEE, for example in Estonia's ICT sector (see Högselius, 2005), is not capable of expanding outwards to capture a slice of the EU or global market, because these innovations are either customized or easy to imitate.
11. See final report of MACROTEC project (Integration of Macroeconomic and S&T Policies for Growth, Employment and Technology) at http://cordis.europa.eu/documents/documentlibrary/82608011EN6.pdf.
12. For example an important issue within this perspective was and still is whether the EU Stability and Growth Pact would have to be relaxed were the EU to come near to achieving the Barcelona target of 3 per cent of R&D in GDP. See von Tunzelmann (2004).
13. 'Editorial, innovations in European and US innovation policy', *Research Policy*, **30**, (2001) 869–72.
14. mckinsey.com/mgi.

REFERENCES

Acemoglu, D., P. Aghion and F. Zilibotti (2002), 'Distance to frontier, selection, and economic growth', *NBER Working Paper Series, WP 9066*, Cambridge, MA: National Bureau of Economic Research.

Aghion, P. and E. Bessonova (2006), 'On entry and growth: theory and evidence', *Revue OFCE*, June.

Aghion, P. and P. Howitt (1997), *Endogenous Growth Theory*, Harvard, MA: MIT Press.

Aghion, P. and P. Howitt (2005), 'Appropriate growth policy: a unifying framework', 2005 Joseph Schumpeter Lecture, delivered to the 20th Annual Congress of the European Economic Association, Amsterdam, 25 August.

Aghion, P. and P. Howitt (2006), 'Joseph Schumpeter Lecture – Appropriate growth policy: a unifying framework', *Journal of the European Economic Association*, **4**(2–3), 269–314.

Aghion, P., W. Carlin and M. Schaffer (2002), 'Competition, innovation and growth in transition: exploring the interactions between policies', *William Davidson Institute Working Paper No. 501*, Ann Arbor, MI: William Davidson Institute, University of Michigan.

Aghion, P., P.A. David and D. Foray (2006), 'Linking policy research and practice in "STIG systems": many obstacles, but some ways forward', *SIEPR Discussion Paper,* Stanford, CA: Stanford Institute for Economic Policy Research.

Andersen, E.S. (2009), *Schumpeter's Evolutionary Economics: A Theoretical, Historical and Statistical Analysis of the Engine of Capitalism*, London: Anthem Press.

Arnold, E. (2004), 'Evaluating research and innovation policy: a systems world needs systems evaluations', *Research Evaluation*, **13**(1), 3–17.

Balaz, V. (2008), 'Annual INNO-policy trendchart report for Slovakia', available at http://www.proinno-europe.eu/node/20531, accessed 10 June 2010.

Barca, F. (2009), 'An agenda for a reformed cohesion policy. A place-based approach to meeting European Union challenges and expectations', Independent Report prepared at the request of Danuta Hübner, Commissioner for Regional Policy, EC, DG Regio, April.

Bernard, F. (2007), 'Innovative enterprises and the use of patents and other intellectual property rights, patents and community innovation survey (CIS) statistics', *Statistics in Focus*, Science and Technology, No. 91/2007, Eurostat, Luxembourg.

Bhidé, A. (2008), *The Venturesome Economy: How Innovation Sustains Prosperity in a More Connected World*, Princeton, NJ: Princeton University Press.

Bruszt, L. and G.A. McDermott (2008), 'Transnational integration regimes as development programs', European University Institute, mimeo.

Bucar, M. (2009), 'INNO-Policy TrendChart: Policy Trends and Appraisal Report: Slovenia 2008', available at www.proinno-europe.eu/. . ./country reports/Country_Report_Slovenia_2008.pdf, accessed 7 August 2010.

Carlin, W. and P.A. Seabright (2007), 'Bring me sunshine: which parts of the business climate should public policy try to fix?', paper presented at the Annual Bank Conference on Development Economics, Bled, Slovenia, 17–18 May.

Castellacci, F. (2006), 'Innovation, diffusion and catching up in the fifth long wave', *Futures*, **38**, 841–63.

Cimoli, M., G. Dosi, R. Nelson and J. Stiglitz (2007), 'Policies and institutional engineering in developing economies', GLOBELICS *Working Paper, No. 07-04*.

Collins, P. and D. Pontikakis (2006), 'Innovation systems in the European periphery: the case of Ireland and Greece', *Science and Public Policy*, **33**(10), 757–69.

Damijan, J., M. Rojec, B. Majcen and M. Knell (2008), 'Impact of firm heterogeneity on direct and spillover effects of FDI: micro evidence from ten transition countries', *LICOS Discussion Paper 218/2008*, available at http://www.econ.kuleuven.ac.be/licos/DP/DP2008/DP218.pdf, accessed 13 August 2010.

Dantas, E., E. Giuliani and A. Marin (2007), 'The persistence of "capabilities" as a central issue in industrialization strategies: how they relate to MNC spill-overs, industrial clusters and knowledge networks', *Asian Journal of Technology Innovation*, **15**(2), 19–43.

Edler, J. and L. Georghiou (2007), 'Public procurement and innovation: resurrect-ing the demand side', *Research Policy*, **36**, 949–63.

Edquist, C. (2002), 'Public technology procurement as an example of public–private partnership', memo written for the Expert Group on Improving the Effectiveness of Direct Support Measures (Direct Measures) to Stimulate Private Investment in Research, mimeo.

Edquist, C. and L. Hommen (1999), 'Systems of innovation: theory and policy for the demand side', *Technology in Society*, **21**, 63–79.

Eliasson, G., D. Johannson and E. Taymaz (2004), 'Simulating new economy', *Structural Change and Economic Dynamics*, **15**, 289–314.

European Bank for Reconstruction and Development (EBRD) (2008), *Transition Report 2008*, London: EBRD.

European Commission (EC) (2004), 'The EU economy: 2004 review', *European Economy No 6. 2004*, Luxembourg: Office for Official Publications of the European Commission.

European Commission (EC) (2009a), 'Five years of an enlarged EU. Economic achievements and challenges', *European Economy*, 1/2009, Brussels: European Commission.

European Commission (EC) (2009b), *European Innovation Scoreboard*, Luxembourg: European Commission.

European Commission (EC) (2009c), *The Role of Community Research Policy in the Knowledge Based Economy*, Expert Group Report, October, Luxembourg: European Commission.

European Union (EU) (2002), 'Innovation tomorrow, innovation policy and the regulatory framework: making innovation an integral part of the broader struc-tural agenda', available at http://cordis.europa.eu/innovation-policy/studies/gen_study7.htm, accessed 8 August 2010.

Featherstone, K. and G. Kazamias (eds) (2001), *Europeanization and the Southern Periphery,* London: Frank Cass.

Freeman, C. and C. Perez (1988), 'Structural crises of adjustment, business cycles and investment behaviour', in G. Dosi, C. Freeman, R. Nelson, G. Silverberg and L. Soete (eds), *Technical Change and Economic Theory*, London and New York: Pinter Publishers, pp. 38–66.

Gaffard, J.-L. and M. Quéré (2006), 'What's the aim for competition policy: optimizing market structure or encouraging innovative behaviors?', *Journal of Evolutionary Economics*, **16**, 175–87.

Georghiou, L. (2006), *Effective Innovation Policies for Europe: The Missing Demand-side*, Helsinki: Prime Minister's Office, Economic Council of Finland.

Gianella, C. and W. Tompson (2007), 'Stimulating innovation in Russia: the role of institutions and policies', *OECD Economics Department Working Papers, No. 539*, Paris: OECD.

Goldberg, I., L. Branstetter, L.J.G. Goddard and S. Kuriakose (2008), 'Globalization and technology absorption in Europe and Central Asia. The role of trade, FDI, and cross-border knowledge flows', *World Bank Working Paper No. 150*, Washington, DC: World Bank.

Hausman, R., J. Hwang and D. Rodrik (2005), 'What you export matters',

Centre for International Development Working Paper No. 123, Cambridge, MA: Harvard University.

Hausman, R., L. Pritchett and D. Rodrik (2004), 'Growth accelerations', *NBER Working Paper Series No. 10566*, Cambridge, MA: National Bureau of Economic Research.

Helpman, E. (ed.) (1998), *General Purpose Technologies and Economic Growth*, Cambridge, MA: MIT Press.

Högselius, P. (2005), *The Dynamics of Innovation in Eastern Europe: Lessons from Estonia*, Cheltenham, UK and Northampton, MA, USA: Edward Elgar Publishing.

Jindra, B. (2005), 'Empirical studies: approaches, methodological problems and findings', in J. Stephan (ed.), *Technology Transfer via Foreign Direct Investment in Central and Eastern Europe*, Basingstoke: Palgrave Macmillan, pp. 30–71.

Kaiser, R. and M. Kripp (2010), 'Demand-orientation in NSI: a critical review of current European innovation policy concepts', paper presented at the Summer DRUID conference, Imperial College London, 16–18 June.

Kaiser, R. and H. Prange (2004), 'Managing diversity in a system of multi-level governance: the open method of coordination in innovation policy', *Journal of European Public Policy*, **11**(2), 249–66.

Knell, M. (2008), 'Embodied technology diffusion and intersectoral linkages in Europe', Europe INNOVA Sectoral Innovation Watch deliverable WP4, Brussels: European Commission.

Kozlowski, J., S. Radosevic and D. Ircha (1999), 'History matters: the inherited disciplinary structure of the post-communist science in countries of central and eastern Europe and its restructuring', *Scientometrics*, **45**(1), 137–66.

Kravtsova, V. and S. Radosevic (2009), 'Are systems of innovation in Eastern Europe efficient?', *UCL Centre for Comparative Economics Working Paper No. 101*, London: University College London, available at http://www.ssees.ucl.ac.uk/economic.htm, accessed 8 August 2010.

Kutlaca, D. and S. Radosevic (2006), 'Macroeconomic regimes and changing cost structures in Central and Eastern Europe', *Acta Oeconomica*, **56**(1), 45–70.

Lundvall, B.-Å. (ed.) (1992), *National Systems of Innovation: Towards a Theory of Innovation and Interactive Learning*, London: Pinter.

Majcen, B., S. Radosevic and M. Rojec (2009), 'Nature and determinants of productivity growth of foreign subsidiaries in Central and East European countries', *Economic Systems*, **33**, 168–84.

McGowan, F., S. Radosevic and N. von Tunzelmann (eds) (2004), *The Emerging Industrial Structure of the Wider Europe*, London: Routledge.

Nauwelaers, C. (2009), 'Policy mixes for R&D in Europe', a study commissioned by the European Commission – Directorate-General for Research, UNU-MERIT, April.

Negro, S. and M. Hekkert (2010), 'Seven typical failures that hamper the diffusion of sustainable energy technologies', paper presented at the Summer DRUID Conference, Imperial College, London, 16–18 June.

Observatory for Science and Technology (OST) (2004), 'Les systèmes nationaux de recherche et d'innovation du monde et leurs relations avec la France. Eléments de rétrospective, situation actuelle et futurs possibles', study prepared by OST, Paris, September.

Organisation for Economic Cooperation and Development (OECD) (2005a), *Governance of Innovation Systems Vol. 1: Synthesis Report*, Paris: OECD.

Organisation for Economic Cooperation and Development (OECD) (2005b), *Business Clusters. Promoting Enterprise in Central and Eastern Europe*, Local Economic and Employment Development Division, Paris: OECD.

Palmade, V. (2005), 'Industry level analysis: the way to identify the binding constraints to economic growth', *World Bank Policy Research Working Paper 3551*, Washington, DC: World Bank.

Perez, C. (1983), 'Structural change and the assimilation of new technologies in the economic and social system', *Futures*, **15**(5), 357–75.

Perez, C. (2000), 'Change of paradigm in science and technology policy', *Cooperation South*, TCDC-UNDP, No. 1-2000, pp. 43–8.

Perez, C. (2001), 'Technological change and opportunities for development as a moving target', *Cepal Review*, **75**(December), 109–30.

Perez, C. (2006), 'Respecialisation and the deployment of the ICT paradigm: an essay on the present challenges of globalisation', *Technical Report EUR22353EN*, IPTS Joint Research Centre, Directorate General, European Commission, available at http://carlotaperez.org/papers/respecialisationandICTparadigm.html, accessed 8 August 2010.

Perez, C. (2007), 'Finance and technical change: a long-term view', in H. Hanusch and A. Pyka (eds), *The Elgar Companion to Neo-Schumpeterian Economics*, Cheltenham, UK and Northampton, MA, USA: Edward Elgar, pp. 820–39.

Perez, C. and L. Soete (1988), 'Catching up in technology: entry barriers and windows of opportunity', in G. Dosi, C. Freeman, R. Nelson and L. Soete (eds), *Technical Change and Economic Theory*, London: Frances Pinter, pp. 458–79.

Possas, M.L. and H.L. Borges (2005), 'Competition policy and industrial development', Institute of Economics, Federal University of Rio de Janeiro, Industrial Policy Task Force.

Radosevic, S. (1999a), *International Technology Transfer and 'Catch Up' in Economic Development*, Cheltenham, UK and Northampton, MA, USA: Edward Elgar.

Radosevic, S. (1999b), 'Transformation of S&T systems into systems of innovation in Central and Eastern Europe: the emerging patterns of recombination, path-dependency and change', *Structural Change and Economic Dynamics*, **10**, 277–320.

Radosevic, S. (2004a), 'What future for S&T in the CEECs in the 21st century', in W. Meske (ed.), *From System Transformation to European Integration. Science and Technology in Central and Eastern Europe at the Beginning of the 21st Century*, Munster: LIT Verlag, pp. 443–78.

Radosevic, S. (2004b), 'Two-tier or multi-tier Europe?: Assessing the innovation capacities of central and east European countries in the enlarged EU', *Journal of Common Market Studies*, **42**(3), 641–66.

Radosevic, S. (2004c), '(Mis)match between demand and supply for technology: innovation, R&D and growth issues in countries of central and eastern Europe', in W.L. Filho (ed.), *Supporting the Development of R&D and the Innovation Potential of Post-Socialist Countries*, Amsterdam: IOS Press, pp. 71–82.

Radosevic, S. (2008), 'Science–industry links in CEE and CIS: conventional policy wisdoms facing reality', paper presented at PRIME Globelics Conference, Mexico City, 24 September.

Radosevic, S. (2009a), 'Policies for promoting technological catching up: towards a post-Washington approach', *Journal of Institutions and Economies*, **1**(1), 22–51.

Radosevic, S. (2009b), 'Research and development and competitiveness, and European integration of South Eastern Europe', *Euro-Asia Studies*, **61**(4), 621–50.

Radosevic, S. and L. Auriol (1999), 'Patterns of restructuring in research, development and innovation activities in Central and Eastern European countries: analysis based on S&T indicators', *Research Policy*, **28**, 351–76.

Radosevic, S. and B. Lepori (2009), 'Public research funding systems in central and eastern Europe: between excellence and relevance: introduction to special section', *Science and Public Policy*, **36**(9), November, 659–66.

Radosevic, S. and F. Sachwald (2005), 'Is enlargement concealing globalization? Location issues in Europe', *Notes de l'Ifri 58*, Paris: Institut Français des Relations Internationales.

Radosevic, S., M. Savic and R. Woodward (2010), 'Knowledge-based entrepreneurship in Central and Eastern Europe: results of a firm level based survey', in F. Malerba (ed.), *Knowledge-Intensive Entrepreneurship and Innovation Systems. Evidence from Europe*, London and New York: Routledge, pp. 198–218.

Rodrik, D. (2008), 'Normalizing industrial policy, commission on growth and development (Spence Report)', *Working Paper No. 3*, Washington, DC: World Bank.

Sass, M., M. Szanyi, P. Csizmadia, M. Illésy, I. Iwasaki and C. Makó (2009), 'Clusters and the development of supplier networks for transnational companies', *Working Paper No. 187*, Budapest: Institute for World Economics, Hungarian Academy of Sciences.

Technopolis (2006), 'Strategic evaluation of innovation and knowledge in the structural funds', Brussels: European Commission, Directorate-General for Regional Policy, available at: http://ec.europa.eu/regional_policy/sources/docgener/evaluation/pdf/strategic_innov.pdf, accessed 24 June 2010.

Teece, J.D. (1992), 'Competition, cooperation, and innovation. Organizational arrangements for regimes of rapid technological progress', *Journal of Economic Behavior and Organization*, **18**, 1–25.

Varblane, U. (ed.) (2008), 'The Estonian economy: current status of competitiveness and future outlooks. Estonia in Focus', Estonian Development Fund, Tallinn.

Veblen, T. (1915), *Imperial Germany and the Industrial Revolution*, New York: Macmillan.

von Hippel, E. (1986), 'Lead users: a source of novel product concepts', *Management Science*, **32**(7), 791–805.

von Tunzelmann, N. (2004), 'Integrating economic policy and technology policy in the EU', *Revue d'Economie Industrielle*, **105**, 85–104.

World Bank (2010), *EU10 Regular Economic Report: Safeguarding Recovery*, Washington, DC: World Bank.

World Economic Forum (WEF) (2009), *The Global Competitiveness Report 2009–2010*, Geneva: WEF.

3. Fostering growth in CEE countries: a country-tailored approach to growth policy

Philippe Aghion, Heike Harmgart and Natalia Weisshaar

3.1 INTRODUCTION

Transition countries, particularly those in Central and Eastern Europe (CEE), have embarked on strong and sustainable growth paths. This chapter provides a general framework for designing medium-term, growth-enhancing policies, while focusing on specific policies relating to competition, education and finance.

Since the late 1980s the Washington Consensus has generally been the prevailing view about which policies are most conducive to good growth performance. This view asserts that, no matter what the country's geographical location or current level of development, the appropriate policy package to achieve growth is to liberalize trade and competition, privatize state-owned firms and maintain a stable macroeconomic environment. In addition, the Consensus highlights the importance of property rights protection and enforcement of contracts as essential preconditions for entrepreneurship and growth to flourish. However, this view is beginning to be challenged. For example, it has been argued that countries in South-East Asia[1] have grown rapidly since the 1970s without fully liberalizing trade, while China is making huge economic strides without privatizing its large state enterprises.

A report commissioned by the World Bank, known as the Spence Report,[2] puts forward policy recommendations that are more nuanced than the Washington Consensus and take account of the different circumstances of countries or regions. It emphasizes the common role of education, trade, competition and labour market mobility in fostering growth across a wide range of countries, and stresses the importance of government commitment to pursuing growth-enhancing policies in the long term.

Policies such as those advocated in the Spence Report are likely to foster long-term growth in the transition countries, although there are several crucial factors that must be kept in mind. The different regions – Central Eastern Europe and the Baltic states (CEB), South-Eastern Europe (SEE) and the CIS+M – had very different starting points in transition in terms of income, education and infrastructure. They were substantially different in the extent to which they were integrated into the European Union (EU) and had been able to develop market institutions. There were also differences in dependency on natural resources. Countries that rely heavily on natural resources tend to suffer from high exchange rates that reduce the scope for economic diversification. More importantly, and particularly when resource-rich countries are non-democratic, they tend also to have higher levels of corruption, poorer governance, and spending priorities that can adversely affect growth.

Education and competition are also of key importance for growth in transition countries. Aside from being among the main policy areas considered by the Spence Report, there is an extensive empirical literature showing that both education and competition matter for growth, even taking account of institutions (Aghion and Howitt, 2006). The potential growth-enhancing effect of education in particular, has been studied in depth using datasets with large numbers of countries (Barro and Sala-i-Martin, 1995), and the growth-enhancing effect of competition has been emphasized in recent cross-country and cross-sector analyses (Aghion, Bloom et al., 2005). These are also areas where substantial progress is needed to catch up with average OECD levels, and which clearly are susceptible to changes in policy preferences and design.

3.2   OVERVIEW OF GROWTH IN TRANSITION COUNTRIES

Table 3.1 presents the growth experience of the three transition sub-regions, the OECD countries and some selected benchmark countries. Growth in the transition economies is substantially higher than in the euro zone, and is above the world average rate. This is because they are middle-income countries that are catching up with economies that are more advanced in respect of capital investment and knowledge acquisition. Their growth is fast because in general it is easier to imitate technologies that have been pioneered elsewhere than to innovate. Also, capital accumulation involves decreasing returns: for countries with already accumulated capital, increasing this stock raises output by more than in countries that have little accumulated capital. Growth rates in the CIS+M have been very high due mainly to the rising price of energy since the beginning of the 2000s.

Table 3.1   Level of GDP per capita and average annual growth rates

Country groups	Indicator	1991	1996	2000	2003	2007
CEB	GDP per capita (real)*	11,026.0	10,859.7	12,835.2	14,722.4	19,030.2
	3 yr av. annual growth in %	−4.1	4.2	4.8	6.8	
	6 yr av. annual growth in %	0.3	4.4	5.8		
SEE	GDP per capita (real)*	6,688.3	5,980.1	6,607.5	7,475.2	9,107.9
	3 yr av. annual growth in %	−1.9	4.4	3.7	5.9	
	6 yr av. annual growth in %	−0.2	4.2	4.8		
CIS, non-resource-rich**	GDP per capita (real)*	3,875.7	2,139.1	2,514.9	3,127.2	4,449.8
	3 yr av. annual growth in %	−16.4	3.2	7.3	8.1	
	6 yr av. annual growth in %	−8.0	4.5	7.7		
CIS, resource-rich***	GDP per capita (real)*	6,231.5	3,884.5	4,548.6	5,696.0	7,895.8
	3 yr av. annual growth in %	−12.2	3.5	7.8	11.9	
	6 yr av. annual growth in %	−6.6	5.8	9.9		
Non-OECD****	GDP per capita (real)*	5,380.1	6,018.2	6,358.6	7,123.6	7,385.1
	3 yr av. annual growth in %	2.8	4.1	3.5	5.5	
	6 yr av. annual growth in %	3.9	3.8	4.4		
OECD****	GDP per capita (real)*	23,639.8	25,614.6	29,206.5	30,306.2	34,084.4
	3 yr av. annual growth in %	2.1	3.7	2.0	3.3	
	6 yr av. annual growth in %	2.8	3.3	2.6		

Notes:
*   *real GDP per capita based on ppp 2005 international $ (Source: WDI, 2009)*
**   *CIS non-resource-rich: Armenia, Belarus, Georgia, Kyrgyz Republic, Moldova, Mongolia, Tajikistan, Ukraine*
***   *CIS resource rich: Azerbaijan, Kazakhstan, Russian Federation, Turkmenistan, Uzbekistan*
****   *excluding transition countries*

Source: World Development Indicators (2009), authors' calculations.

In relative terms, real GDP per capita is highest in CEB, followed by SEE, and then the resource-rich CIS+M, with the non-resource-rich Commonwealth of Independent States countries in the bottom of the ranking. In CEB, per capita GDP is still only 55 per cent of the OECD average and in the non-resource-rich Commonwealth of Independent States countries it is 12 per cent of the OECD average. It is evident that GDP growth has been highest in those countries with the lowest levels of GDP, an indication of the scope for these economies to catch up.

It should be noted that growth rates in the transition economies, including the CIS+M, are well below the 8–10 per cent range experienced by China and India, and are more in line with the second wave of emerging Asian economies such as Indonesia, Malaysia or Thailand.

3.3   FRAMEWORK FOR DESIGNING GROWTH POLICIES

The starting point for analysing growth at country level generally has been to view the flow of domestic output as generated by a given stock of factors of production, particularly capital and labour and their respective levels of productivity. A country with limited capital can grow faster by accumulating more capital, whereas a country with already accumulated capital achieves less gain from increasing its rate of accumulation. Eventually, accumulating more capital will entail more capital depreciation than can be generated in terms of added output.

3.3.1   From Neoclassical to New Growth Theory

A natural starting point is to specify an aggregate production function which describes how domestic output flow is generated from a given stock of production factors, in particular capital. This production typically involves decreasing returns to capital accumulation, that is: one more unit of capital yields less additional output, the greater is the already accumulated capital. For example, we can write:

$$Y = AK^{\alpha},$$

where K is the capital stock, A is a productivity factor that reflects the existing stock of knowledge and the resulting efficiency of capital and labour in producing the final output, and $\alpha < 1$ so that the production technology exhibits decreasing returns to capital accumulation.

Growth in output can be the result: (i) of the accumulation of capital K; (ii) of increases in the productivity factor A, that is, from productivity growth. Neoclassical growth theory emphasizes capital accumulation as the main source of growth, thus taking productivity A as fixed. Capital accumulates over time as a result of investment and capital depreciation.

$$\dot{K} = I - \delta K,$$

where K_{+1} is the capital stock for the next period, so that $K_{+1} - K$ is the net increase in capital stock per unit of time and δ is the rate of depreciation of capital.

Using the fact that investment is equal to aggregate savings in equilibrium, and assuming that people save a constant fraction s of their income, we have

$$I = sY,$$

so that the growth rate of capital is given by:

$$g = \frac{K_{+1} - K}{K} = sY/K - \delta = sAK^{\alpha-1} - \delta.$$

In particular, a country with little capital endowment K grows fast by accumulating more capital, whereas a country that has already accumulated capital does not gain much by increasing its rate of capital accumulation because of the decreasing returns to capital accumulation. And, as a result of these decreasing returns, in the long run, growth simply stops. This happens when depreciation catches up to savings, that is, when enough capital has been accumulated so that $sAK^{\alpha-1} \leq \delta$.

3.3.2 The Contribution of New Growth Theory

New growth theory explains sustained long-run growth by endogenizing productivity growth, that is, growth in A. The idea is that productivity growth results from innovations. Schumpeterian growth theory, in particular, emphasizes the role of quality-improving innovations, that is, innovations that increase the productivity A of production factors. New growth theory recently (for more detail see Aghion and Howitt, 2006) has linked productivity growth to innovation and, in its turn, innovation is motivated by the prospect of the above-normal returns that successful innovators can realize. The theory suggests that innovation and, therefore, productivity growth, will always be fostered by:

- better protection of intellectual property rights to allow successful innovators to benefit from their endeavours;
- financial development, since tight credit constraints will limit the ability of entrepreneurs to finance new innovative projects;
- better education, since this will improve the ability to innovate in and/or imitate leading edge technologies;
- macroeconomic stability, which allows for a lower, risk-adjusted interest rate that will enable entrepreneurs to invest more in growth-enhancing projects.

Another important feature of innovation is what Joseph Schumpeter refers to as 'creative destruction', that is, that innovations displace old products or old technologies. Therefore, faster growth typically implies a higher rate of firm turnover, as the process of creative destruction generates the entry of new innovators and the exit of old ones. Indeed, competition is likely to enhance growth because it enables this process of turnover.

Overall research and development (R&D) spending, and the number of patent registrations[3] are good indicators of a country's level of innovative activity. The transition countries lag well behind the OECD average. Also, although income levels may be catching up, this is not the case for investment in innovation. In contrast to emerging Asia, where R&D has been increasing, the transition countries have not raised their innovative activity levels (see EBRD, 2008).

More formally, Schumpeterian theory begins with a production function specified at industry level:

$$Y_{it} = A_{it}^{1-\alpha} K_{it}^{\alpha}, 0 < \alpha < 1$$

where A_{it} is a productivity parameter attached to the most recent technology used in industry i at time t. In this equation, K_{it} represents the flow of a unique intermediate product used in this sector, each unit of which is produced one-to-one by capital. Aggregate output is the sum of the industry-specific outputs Y_{it}.

Each intermediate product is produced and sold exclusively by the most recent innovator. A successful innovator in sector i improves the technology parameter A_{it} and, thus, is able to displace the previous innovator and be the incumbent intermediate monopolist in that sector until displacement by the next innovator. Thus, the first implication that distinguishes the Schumpeterian Paradigm from the AK and product-variety models is that *faster growth generally implies a higher rate of firm turnover, because this process of creative destruction generates the entry of new innovators and the exit of former innovators.*

3.4 DISTANCE FROM THE TECHNOLOGY FRONTIER AND THE CHOICE BETWEEN INNOVATION AND IMITATION

There are several ways that a country can increase its productivity growth. One is to imitate more advanced technologies invented elsewhere; another is to produce a leading-edge domestic innovation that builds on and extends international technology standards. A country that is far from the technological frontier[4] can make substantial productivity growth leaps by imitating leading technologies developed elsewhere. However, countries closer to the technological frontier will need to rely primarily on new innovations, which are more difficult to generate, in order to grow further.

More formally, a frontier innovation leapfrogs the best technology available before the innovation, resulting in a new technology parameter A_i in the innovating sector i, which is some multiple γ of its pre-existing value. An imitation is a technological activity that allows the country or sector to catch up to the international technology frontier $\overline{A}_t$ which typically we take to represent the stock of global technological knowledge available to innovators in all sectors of all countries. In the former case the country is making a leading-edge innovation that builds on and improves the leading-edge technology in its industry. In the latter case the innovation is an imitation of a technology that has been developed elsewhere.

For example, consider a country where, in any sector, leading-edge innovations take place at the frequency u_n and implementation innovations take place at the frequency u_m. Then the change in the economy's aggregate productivity parameter A_t between time t and time $t + 1$ will be:

$$A_{t+1} - A_t = u_n(\gamma - 1)A_t + u_m(\overline{A}_t - A_t)$$

and hence the country's growth rate will be:

$$g_t = \frac{A_{t+1} - A_t}{A_t} = u_n(\gamma - 1) + u_m(a_t^{-1} - 1)$$

where:

$$a_t = A_t/\overline{A}_t$$

is an inverse measure of the distance to the frontier.

In particular we see immediately that a country that lies farther away from the world technology frontier (that is, has a lower a_t), should grow faster, everything else remaining equal, since it will make bigger leaps each time it imitates the leading technology. In other words, it benefits from higher knowledge spillovers from the more advanced countries.

We could take as given the critical innovation frequencies u_m and u_n which determine a country's growth path. However, Schumpeterian theory derives these innovation frequencies endogenously from the profit-maximization problem facing the prospective innovator. In equilibrium, these frequencies typically depend upon the economy's institutional characteristics, such as property rights protection and financial system, and on government policy; moreover, this dependence in terms of impact and mix of innovation may vary with the country's distance to the technological frontier a.

In addition, the above growth equation naturally captures Gerschenkron's idea of 'appropriate institutions'. Suppose, for instance, that the institutions that favour imitation (that is, that lead to firms emphasizing u_m at the expense of u_n) are not the same as those that favour leading-edge innovation (that is, that encourage firms to focus on u_n): then, a country far from the frontier will be able to maximize growth by favouring institutions that facilitate imitation. However, as it catches up to the technological frontier, in order to sustain this high growth rate its institutions will have to shift their focus from imitation-enhancing to innovation-enhancing as the relative importance of μ_n for growth is also increasing.

Historically, the main examples of imitation of existing technologies are Japan and Europe after 1945, and more recently the economies constituting the so-called Asian Tigers, such as China, Korea and Chinese Taipei. Imitation has tended to emerge when:

- large firms can take advantage of economies of scale;
- there is limited labour mobility among firms, so that workers' skills remain largely specific to their firms;
- there is limited competition and entry, allowing large firms to survive longer and make long-term investments in capital and labour;
- financial systems can provide long-term bank finance.

In contrast, countries that innovate at the technological frontier have tended to require:

- high labour market mobility, so that innovating firms entering new markets are able to find workers whose skills match their needs;
- more intense product market competition and low entry barriers;
- a focus on tertiary and, particularly, graduate education, and universities that produce researchers and generate the basic science that firms need to harness in order to innovate;

- a bigger role for non-bank finance and stock markets to help in the selection of the most promising innovative projects to finance.

The overall effect of competition or entry on growth at country level, therefore, depends on the proximity of local industries to their respective technological frontiers, and on the technological level of new entrants.

In light of these distinctions and differences, what can be said about where the transition countries lie in relation to the technological frontier, and how much they have caught up with it since the early 1990s? Although the transition region as a whole is lagging far behind the frontier – in the range of 12–42 per cent of US labour productivity levels – the gap has been narrowing, and particularly among the EU accession countries (CEB and Bulgaria and Romania). This indicates that there is considerable variation within the transition group of countries.

3.5 POLICIES FOR GROWTH

3.5.1 Competition and Entry

This section focuses on the related issues of competition and market entry and looks at how the framework outlined above can be translated into more specific policies to encourage growth. It provides some evidence regarding the number of market entrants and competition in the transition countries and considers appropriate policy responses.

We argue that product market competition enhances innovation, labour productivity and growth (see Carlin et al., 2004; Aghion et al., 2002). Existing levels of product market competition (measured inversely by profit margins)[5] are significantly below OECD averages. Policies to encourage product market competition, therefore, are likely to have positive pay-offs both for old firms where competition can be a substitute for effective corporate governance, and new firms where these policies spur innovation by increasing the incremental profits arising from being ahead of competitors.

A comparison of entry rates in the transition countries with the OECD average, shows that although still below OECD levels, they have been increasing in the former. This is attributable in part to deterrents in the business environment to new entrants. These barriers range from limits on availability of credit, to levels of taxation and other regulatory impediments.

A recent study found that, particularly in Western Europe, credit constraints are one of the main barriers to the entry and post-entry growth of very small firms (Aghion et al., 2007). This is especially true in transition economies: small firms regularly report access to finance as a major impediment to their business activity and growth potential (see EBRD, 2006).

In relation to competition, evidence from the European Bank for Reconstruction and Development (EBRD) and the World Bank Business Environment and Enterprise Performance Survey (BEEPS), covering more than 20 000 firms in 26 transition countries, gives some indication of the extent of competition, focusing on trade and imports. The BEEPS results show that competition between imported and domestic products has become more intense in CEB and SEE over time, but less intense in the CIS+M. This is due mainly to increased trade integration in the CEB and SEE regions, principally as a result of the EU accession process, but also because of increased intra-regional trade. With the exception of commodity-induced trade in the resource-rich countries, the CIS+M region shows lower internal and external trade integration.

Focusing on manufacturing industries in the sub-regions, SEE and the CIS+M show lower levels of product market competition than the OECD average. Pricing power is an indication of the extent of the competitive pressure in an industry. Pricing power or industry mark-up can be measured directly using the Lerner Index, which measures the difference between value added and the total wage bill expressed as a share of gross output. Using a large UNIDO (United Nations Industrial Development Organization) cross-country/cross-industry dataset to compare average mark-ups across countries between 1998 and 2007 we see that average mark-ups are higher in the transition countries (particularly SEE and the CIS+M) than in the OECD countries, indicating that competition is less intense. The evolution of mark-ups reveals that they have declined over time throughout the CEE region, and that mark-ups in the CIS+M and the SEE countries are generally higher than in the OECD and CEB countries, and higher than the world average. This can be explained by a number of factors, including regulated product markets as well as low levels of diversification away from commodities in the resource-rich CIS+M. Mark-ups remain substantially higher in the transition countries, particularly the CIS+M and SEE. Reducing them through the effect of increased foreign competition, either from further trade integration or direct entry of foreign firms, would help to boost innovation and productivity growth.[6]

Figure 3.1, based on BEEPS data, depicts the impact of competition on product innovation. It shows that domestic and foreign competition have a larger impact on the non-resource-rich countries than on the resource-rich countries and, also, that pressure tends to come from domestic rather than foreign competition. These findings may reflect the fact that:

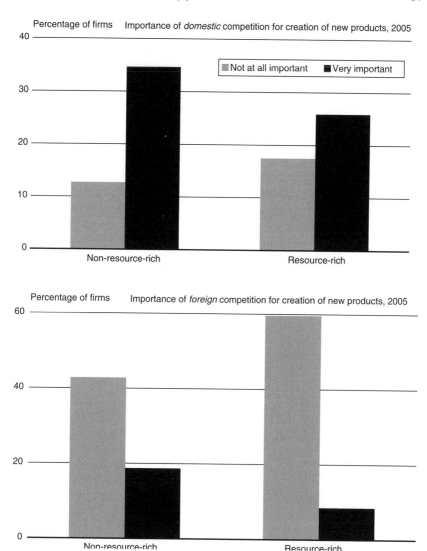

Note: The figures show the proportion of firms claiming that domestic or foreign competition is a (i) not at all important or (ii) very important pressure for the creation of new products.

Source: BEEPS (EBRD-World Bank, 2005).

Figure 3.1 Importance of competition for the innovation of new products

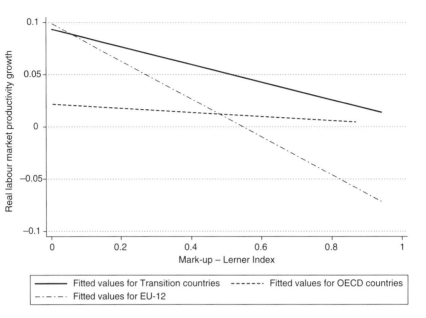

Note: The figure depicts predicted values from regression results of labour productivity growth on the Lerner Index, Lerner Index squared, accounting for year, industry and countries. Regressions were run separately for OECD and EU-accession countries and transition countries.

Source: UNIDO and Vienna Institute for International Economic Studies WIIW industry statistics and authors' calculations.

Figure 3.2   Labour productivity growth and manufacturing mark-ups

(i) although foreign competitors might be a greater challenge for local incumbents in terms of technological sophistication, most firms compete mainly with their domestic counterparts (which may reflect a lack of trade integration and additional barriers to entry for foreign firms); and (ii) firms in resource-rich countries are more likely to benefit from government subsidies that shield them from some of the effects of competition.

Recent evidence (see Nickell, 1996; Blundell et al., 1999; Aghion, Bloom et al., 2005) indicates that product market competition and entry have positive and significant effects on productivity growth in emerging market economies such as India and South Africa. It appears that this holds also for transition countries. Figure 3.2 shows that lower competitive pressure is indeed associated with lower productivity growth, and the relationship is substantially stronger for the EU accession countries and the group of transition countries as a whole, compared to the OECD economies (as can be seen by the relative slope of the three lines in Figure 3.2). Therefore, although the transition countries are not very near the world technology frontier, they are sufficiently close for competition to be growth-enhancing.

This analysis confirms that lack of competition leads to less intense innovation, which, in turn, slows the speed at which productivity catches up to the technological frontier. Also, too little product market competition directly affects labour productivity growth. This is true for countries that are close to the frontier – such as the OECD group – and also for transition countries.

In light of this discussion, the question that emerges is how transition countries can ensure that there is an appropriate level of product market competition and that this, in turn, translates into labour market productivity and, ultimately, overall economic growth. There is an important role here for effective institutions, such as competition authorities. In a 2007 competition survey, the EBRD measured the efficiency and effectiveness of these institutions: overall expenditure on competition regulation and enforcement as a share of GDP; and an index for the efficiency of enforcement focusing on competition authority decisions related to market dominance. With the exception of spending, the indicators are higher for the CEB countries than for the rest of the transition region. This is being driven partly by the process of EU accession of the CEB region and some SEE countries, which has introduced standardized competition legislation and enforcement procedures. SEE is lagging behind CEB in terms of contract enforcement processes (delays being due primarily to case backlog and an insufficient number of judges), although overall expenditure is high.

Experience from other emerging countries shows that independent and transparent competition authorities can exercise a positive influence on product market competition. Rather than narrowly focusing on curbing the dominant firms in the market, competition authorities need to employ an approach that keeps entry and exit barriers low and provides incentives to firms to innovate. This means adopting a broad-based approach with deregulation at its core to ensure clear and quick licensing procedures which eliminate bureaucratic bottlenecks. For registration, a one-stop shop system and/or (ideally) on-line registration would significantly reduce the numbers of transactions and the time spent on related processes. Applicants could submit a single form containing all the information required by the various agencies to a single entity. Although some transition countries have moved towards such a system, registration requirements in the CIS+M group are still cumbersome. There are also large differences across SEE countries in terms of the indicators used to measure the institutional conditions for product market competition (see World Bank, 2008).

Another important element for ensuring ease of entry and subsequent product market competition, is the cost of construction licences and, more generally, property rights, registration and collateralization. Again, there are large differences across countries. For example, to register a property takes between three days in Lithuania, and 331 days in Bosnia and Herzegovina (the OECD average is 28 days), while costs vary from 0.1 per cent of the property value in Georgia, to 11 per cent in Hungary.

In addition to the traditional role of competition authorities of investigating anti-competitive practices by firms, they are also responsible for holding to account and, if necessary, filing cases against, local and regional government bodies that restrict competition. Case-by-case investigations of violations of competition law need to be accompanied by competition advocacy in order to cultivate entrepreneurial activity and provide functional support to new firms.

3.5.2   Education

Research based on a large number of countries and data points for the period 1960–92 shows that education enhances growth.[7] Higher levels of education enhance innovation since a better educated population is also better equipped to contribute. Higher average levels of education are also crucial for successful imitation and faster adaptation of existing modern technologies.

Studies of education and growth often measure education in terms either of spending (the fraction of aggregate GDP devoted to education) or of attainment (proportion of the working age population with particular qualifications). However, these studies have been extended to include measures for the quality of education (see Hanushek and Kimko, 2000; Hanushek and Woessmann, 2008).

Using internationally comparable student test scores to measure quality of their cognitive skills, we can identify a positive and significant correlation between long-term growth and the quality of education, for a large sample of countries (see Figure 3.3).

The same research (using information for 50 countries for the period 1960–2000) finds that countries with better test scores show significantly higher rates of annual growth in GDP per capita. More specifically, an increase of 100 points for the test results[8] is associated with an increase in the rate of annual growth of 1.3 to 2 percentage points. A reform that would improve student outcomes by 50 points, on average would increase GDP by around 5 per cent over a period of 20 years, and by 36 per cent over a period of 75 years (Hanushek and Woessmann, 2008).[9]

Complementary research has analysed the relationship between growth and the composition of education spending (see Aghion et al., 2008).

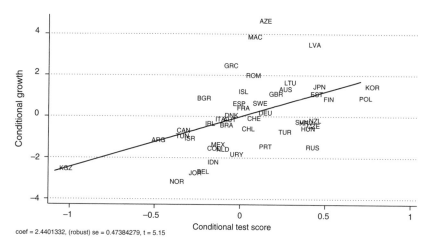

coef = 2.4401332, (robust) se = 0.47384279, t = 5.15

Notes:
PISA 2006 average country scores in reading, mathematics and science (in 100s). The graph shows the effect of an increase of 100 PISA points on long-term growth in per capita GDP (1998–2006), controlling for real GDP per capita in 1998, enrolment rates in higher education (1991), degree of openness to trade and regional differences.
Countries shown on the figure are: ARG – Argentina, AUS – Australia, AUT – Austria, AZE – Azerbaijan, BEL – Belgium, BGR – Bulgaria, BRA – Brazil, CAN – Canada, CHE – Switzerland, CHL – Chile, COL – Colombia, CZE – Czech Republic, DEU – Germany, DNK – Denmark, ESP – Spain, EST – Estonia, FIN – Finland, FRA – France, GBR – United Kingdom, GRC – Greece, HKG – Hong Kong, HRV – Croatia, HUN – Hungary, IDN – Indonesia, IRL – Ireland, ISL – Iceland, ISR – Israel, ITA – Italy, JOR – Jordan, JPN – Japan, KGZ – Kyrgyz Republic, KOR – Rep. Korea, LTU – Lithuania, LUX – Luxembourg, LVA – Latvia, MAC – Macao, MEX – Mexico, NLD – Netherlands, NOR – Norway, NZL – New Zealand, POL – Poland, PRT – Portugal, QAT – Qatar, ROM – Romania, RUS – Russia, SVK – Slovak Republic, SVN – Slovenia, SWE – Sweden, THA – Thailand, TUN – Tunisia, TUR – Turkey, URY – Uruguay.

Source: World Development Indicators 2008 (World Bank, 2008), OECD (2007), authors' own calculations (regression results).

Figure 3.3   Real GDP per capita growth and average PISA 2006 test scores

Results show that the closer a country or region's productivity is to the technological frontier, the more growth-enhancing it becomes to invest in higher education and, particularly, postgraduate education and research. The farther the country or region is from the frontier, the more growth-enhancing it will be to invest in primary, secondary and undergraduate education, which is more likely to make a difference in terms of the country's ability to imitate existing technologies.[10]

However, the complexity of the relationships, and the differences among transition countries, call for a careful, country-based interpretation of results before formulating strong policy recommendations. To suggest that transition economies focus on primary and secondary education simply because they are farther from the frontier would be inappropriate. For example, without a good tertiary education sector, India would have been unable to develop a dynamic service sector. Conversely, while the transition economies can increase their growth potential by investing in higher quality primary and secondary education, this should not be at the expense of undergraduate or even postgraduate education (see also World Bank, 2000; 2005; 2006).

Table 3.2 presents the evolution of education spending and quality in the transition countries based on expenditure and enrolment rates across different groups of transition economies and the OECD, over the period 1999–2006.[11]

Table 3.2 shows that the proportion of expenditure on tertiary education has decreased since 1999 in all transition regions, but has remained fairly constant in the OECD countries. Transition countries spend less per student than the OECD average and have lower enrolment rates. Expenditure per student on primary and secondary education (percentage of per capita GDP) has been generally stable or has increased over the same period, although in the resource-rich CIS+M per student investment decreased between 2003 and 2006. There are large differences across the transition sub-regions: resource-rich countries expend the least on tertiary and primary education, and have much lower tertiary level enrolment rates than non-resource-rich countries. This implies that countries with sharply rising resource flows are failing to exploit these new resources to increase funding for education and, therefore, are possibly missing an opportunity to address some of the shortcomings in their education systems.

Whatever the level of spending as a share of national income, what is important is whether the expenditure and student enrolment numbers are in line with intended education outcomes. One indicator that is comparable over a large set of transition and non-transition countries is the PISA[12] test score, which measures reading, science and mathematics achievements in a standardized way. Figure 3.4 links PISA test scores to education spending, and shows a positive and significant relationship, especially for the transition countries. If we take account of income levels, increased expenditure on education in the transition countries does appear to be associated with better quality education.

However, the quality of education for the transition economies as a whole is well below the OECD average, and in Russia it has even decreased. There are significant differences across the transition regions. While student performance in the CEB countries in 2006 was close to the

Table 3.2 *Expenditure per student at different education levels (as % of per capita GDP) and gross enrolment rates*

Country groups	Indicator	Primary education		Secondary education		Tertiary education	
		1999–2002	2003–2006	1999–2002	2003–2006	1999–2002	2003–2006
CEB	Expenditure/ student	17.4	19.2	21.9	22.9	27.8	24.9
	Gross enrolment rate	101.5	99.5	95.4	98.3	47.4	58.9
SEE	Expenditure/ student	13.0	16.4	17.5	18.7	31.3	26.6
	Gross enrolment rate	100.2	99.7	83.2	87.4	28.3	33.8
CIS, non-resource-rich*	Expenditure/ student	10.1	13.0	12.9	17.4	29.9	26.5
	Gross enrolment rate	102.2	98.8	82.6	87.0	35.3	41.0
CIS, resource-rich**	Expenditure/ student	8.2	8.2	14.7	9.9	14.6	10.0
	Gross enrolment rate	100.6	104.5	86.5	90.9	22.7	34.6
OECD***	Expenditure/ student	18.7	19.7	24.1	25.0	35.8	34.8
	Gross enrolment rate	103.2	102.9	109.8	107.6	54.1	61.6
Selected countries							
Finland	Expenditure/ student	17.6	28.5	25.9	29.4	38.8	37.2
	Gross enrolment rate	100.7	101.0	124.2	118.7	83.6	88.3
France	Expenditure/ student	17.4	17.6	28.2	28.9	29.1	32.2
	Gross enrolment rate	106.2	106.1	109.5	110.4	53.0	55.1
US	Expenditure/ student	19.9	21.6	23.6	25.1	28.0	25.3
	Gross enrolment rate	100.3	98.8	94.0	93.9	73.1	81.8

Notes:
* includes Armenia, Belarus, Georgia, Kyrgyz Rep., Moldova, Mongolia, Tajikistan, Ukraine
** includes Azerbaijan, Kazakhstan, Russian Federation, Turkmenistan, Uzbekistan

Source: World Bank World Development Indicators (2008).

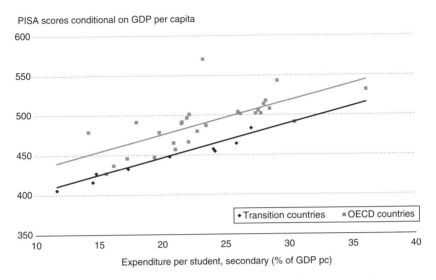

Note: PISA 2006 average country scores in reading, mathematics and science. Graph depicts predicted PISA 2006 results based on a regression of PISA 2006 results on mean expenditure on student (percentage of GDP per capita) 1998–2005 and mean real GDP per capita 1998–2005.

Source: World Development Indicators 2008 (World Bank, 2008), OECD (2007), authors' own calculations.

Figure 3.4   Expenditure per student in secondary education and PISA outcomes

OECD average, in SEE average test scores were quite low, and very low in the CIS+M group of countries. However, compared to countries with similar GDP per capita levels, the transition countries are performing generally better (see EBRD, 2008). Latvia and Poland, for example, achieved substantial improvements in student performance between 2000 and 2006.

The full scope for improvement in cognitive skills in the transition countries becomes evident if we compare results with the OECD and top-scoring comparator countries such as Finland. The mean scores for the PISA tests for mathematics, science and reading skills, for the top performers among the different groups of countries, show that Finland's students achieved the highest score for science with 563.3 points (that is, roughly 50 points above the OECD average). The top performer among the transition countries was Estonia, with average student test scores of 514.6 and 531.4 points for mathematics and science respectively. Russia was the leading country in the CIS+M group and Central Asia, although the performance of Russian students in mathematics and science was below the OECD average. The gaps in the scores for Russia and the overall PISA leaders range from 73.7 points for mathematics, up to 116.1 points for reading skills; the same differences with the best-performing transition country (Estonia) are smaller, but still sizeable at 38.9 and 67.7 points respectively. The considerable gap between the transition countries and the top-performing countries is an indication of the potential for improvements in education quality and, ultimately, the growth potential of the transition region.

The significant differences in the test scores of the transition countries reveal the high potential for future improvements in the quality of cognitive skills. This would have a strong impact on long-term economic growth. Russia, for example, could achieve higher long-term annual GDP growth rates, of between 0.065 and 1 percentage point, merely by catching up to the top PISA performers among the transition countries.

In terms of policy, the transition countries need to invest more overall in education, but in a way that links that investment to better quality. The urgent need is to invest in primary and secondary education, but they need also to improve higher (particularly undergraduate) education. Without such investment, countries will not be able effectively to imitate technological innovations produced elsewhere. At the same time, better monitoring and evaluation systems would increase the effectiveness of investment in education. Further participation in school-based, national and international assessments, such as PISA, will help policy-makers by revealing their countries' relative performance in education (see World Bank, 2006).

Although Figure 3.4 indicates that higher expenditure per student tends to be associated with better student performance, the aggregate results mask considerable differences among countries. Studies analysing the effect of school inputs and resources – typically teacher: student ratios, class sizes, textbook provision, teacher training and experience, monitoring of schools, school facilities and administration – provide mixed evidence on strategies aimed at improving educational outcomes that would apply to all countries. Overall, there needs to be better use and targeting of education spending, improved teacher quality, greater accountability to parents, students and national educational authorities, and stricter adherence to standards. Transparency through public participation and feedback mechanisms is important for effective delivery and regulation of education. In the transition countries, there is a notable lack of consultation. Accountability in the education system could be promoted through decentralization and improvements to local school management practices.

Another issue of concern to policy-makers is related to equal and easy access to education. The background of the student seems to matter greatly for educational performance in the transition countries, and much more than resources or the institutional setting (see Ammermueller et al., 2005). This result, in addition to highlighting this inequality, highlights the need for policy reform to help secure funding and improve access to education (including pre-primary education) for children from less affluent families. Poor regions should receive financial transfers from central government. Sustainability of and equity in the financing of education could be improved through the use of funding formulas based on per student spending. This would help to combat poverty by targeting public education resources at the poor (see World Bank, 2000).

The PISA results show that students in the transition countries are lagging behind the OECD average by around 17 points (and by 65 points behind the top performers) for problem-solving skills and applying knowledge in new areas. Changes in the primary and secondary curricula and vocational education are needed to enhance critical thinking and provide children with more general and more relevant skills. In this context, the secondary education curriculum is crucial, since it must fulfil the dual purpose of linking directly to the labour market and preparing students for tertiary education (see World Bank, 2006). Figure 3.5 shows that the transition countries have a higher proportion of social science graduates at tertiary level than the OECD average and some selected comparator countries (for example Sweden).

In terms of vocational training, the transition countries have inherited a very narrowly defined curriculum which needs to be broadened and updated to increase the relevance of their vocational training programmes. Greater involvement of private business in the design of training programmes will be important.

The wider problem of skills mismatches in the labour market is important and needs to be addressed by policy. The EBRD's Life in Transition Survey (LITS), carried out in 2007, found that one third of all employees was not engaged in work related to educational attainment and formal training.[13] This mismatch is greatest in the CIS+M group. The problem is not unique to the transition countries, but highlights the need for more investment in lifelong learning and retraining, to enable workers and firms continually to upgrade their skills. Providing tax incentives for workers and firms to take up training opportunities has proved generally to be more fruitful than attempts to set up public training programmes.

3.6   FINANCIAL CONSTRAINTS

A review of the literature by Levine (2004) points to the positive effect of financial development on growth. The underlying idea is that firms

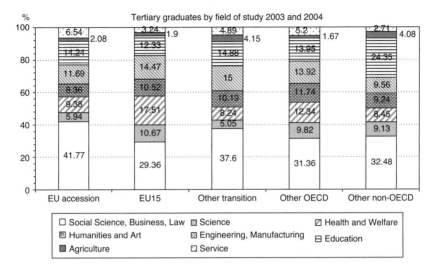

Note: Defined as the number of students graduating in a particular field expressed as a percentage of the total number of graduates of tertiary education.

Source: UNESCO Institute for Statistics.

Figure 3.5 *Tertiary graduates by field of study (in per cent of total graduates)*

face credit constraints due to asymmetric information between financiers and firms, or enforcement problems (borrowers take the money and run unless properly monitored). The lower these constraints, the easier it is for innovating firms to finance their projects and advance knowledge. Levine (2004) provides cross-country panel regressions of growth on two measures of financial development: ratio of bank credit to GDP and degree of stock market capitalization, both of which are positively and significantly related to growth in the long run.

While labour market rigidities are often presented as the main impediment to firms' entry, mobility and post-entry growth, it would seem that financial constraints are at least as important. A study by Aghion, Fally and Scarpetta (2007) shows that credit constraints are the main barrier to the entry and post-entry growth of very small firms, while labour market regulations inhibit the entry of larger firms. This is particularly significant for transition economics since small firms in transition countries regularly report access to finance as a major impediment to their business activity[14] and growth potential, and in these countries the growth of small firms is also significantly positively related to overall growth (see Figure 3.6).

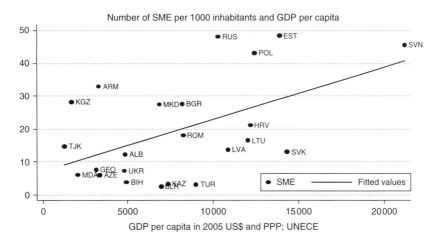

Figure 3.6   SME density and GDP growth

Aghion and Akcigit (2008) use cross-country/cross-industry panel data to show that the closer a local industry is to the world technology frontier, the more that industry benefits in terms of growth, from higher stock market development in the country, while growth in industries that are far behind the frontier is favoured more by the availability of bank credit. Again, this is based on the idea that growth in the more advanced countries relies more on frontier innovations. However, innovation, in its turn, is riskier than imitation activities and typically involves firms with less financial capital for use as collateral in case of repayment default. Equity financing compensates the financier for this added risk through the endowment of a higher share of the returns, even if these are already high.

Figure 3.7 shows the average distances to the technological frontier (measured as the distance in total factor productivity (TFP) as a share of US TFP) for the transition regions, the OECD and the world, and a subset of OECD benchmark countries. The distance from the frontier for manufacturing is still quite large for all the transition regions; while some OECD countries, such as France and the UK, show decreased competitiveness against the US over time, the distance of the transition countries is constant.

This suggests that the transition economies that are less advanced than the OECD countries will be more reliant on bank finance than the latter. Evolution of the private credit to GDP ratios and levels of stock market capitalization for the OECD and transition countries over past decades, underline that the transition countries are significantly behind the OECD and also China.

Unlike competition policy or trade liberalization, financial development

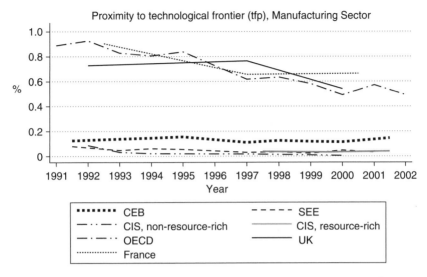

Source: UNIDO and Vienna Institute for International Economic Studies WIIW industry statistics and authors' calculations.

Figure 3.7   TFP frontier for the manufacturing sector

is a gradual process. However, it can be speeded up through the setting up of good banking regulations, elimination of non-performing loans, and opening up of domestic economies to foreign banks and direct investment. In fact, the increase in the share of foreign banks since 1999 has contributed to a decrease in the percentage of non-performing loans in the banking system. This is not surprising: new foreign banks that take over local banks impose stricter prudential regulations while also improving the efficiency of the overall banking system.

We have argued that the relative importance of banks and stock markets depends on the level of development of the country, with more advanced countries relying more on the stock markets to finance their investments. Figure 3.8 depicts the excessive importance in Russia of stock market relative to bank finance, which is due mainly to the relative slowness of banking reform in Russia, which has constrained bank-based finance.

3.7   FINANCING REFORM

Promoting and enforcing competition generally makes greater demands on political will and reform capacity than on government budgets.

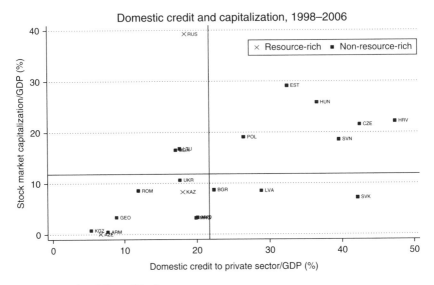

Source: National Central Banks.

Figure 3.8   Domestic credit and stock market capitalization

However, entrenching education reforms that support long-term sustainable growth is more financially taxing. While the private sector has a part to play in overcoming the skills mismatch, governments are crucial for ensuring access to, and the quality of, formal education.

For many transition country governments, the scope for increasing spending on education is limited by their overall debt levels, budget deficits and capacities to tax their citizens – aggravated further by the current ongoing financial crisis. However, compared to the advanced market economies, many transition countries are relatively well placed in terms of their capacity to sustain debt and their ability to repay it. For example, with the exception of Hungary, all CEB transition economies have debt to GDP levels below the 60 per cent benchmark imposed by the EU Maastricht Treaty for joining the economic and monetary union. In addition, most transition countries run relatively small budget deficits, and over 40 per cent have budget surpluses. On the downside, inflation is significantly higher in transition countries than in most OECD countries. Moreover, tax revenues as share of GDP are generally substantially lower than in the OECD countries.

The last point – the ability to tax – requires very careful consideration. A closer look at the BEEPS results for 1999 to 2005 sheds some light on the overall low tax revenue figures in the transition countries (see Table

Table 3.3   Subjective indicators on tax compliance (BEEPS survey), percentage of firms

Year	1999		2002		2005	
Estimated	% of firms					
share of sales reported to the tax authorities	CEB+2	All other transition countries	CEB+2	All other transition countries	CEB+2	All other transition countries
100%	33.64	31.43	56.55	52.28	62.61	67.34
90–100%	13.77	12.55	5.94	2.45	4.30	2.19
80–90%	10.47	12.21	8.85	6.14	8.17	5.16
70–80%	8.24	7.74	10.49	8.65	11.55	7.30
60–70%	5.52	5.07	5.99	6.53	4.59	5.10
50–60%	6.92	10.88	2.46	4.20	2.55	2.79
25–50	6.51	9.50	7.88	12.43	5.39	8.06
≤ 25%	14.92	10.62	1.84	7.32	0.83	2.05
Total	100	100	100	100	100	100

Note: Based on the BEEPS survey question: 'What percentage of the sales of a typical firm in your area of activity would you estimate is reported to the tax authorities, bearing in mind difficulties with complying with taxes and other regulations?' (BEEPS 1999). Although the exact wording of this question has been slightly modified over time, the answers are comparable over time.

Source: BEEPS (EBRD-World Bank, 1999, 2002, 2005).

3.3). Over time, the developments regarding tax compliance seem to be in the right direction: the estimated percentage of firms underreporting sales (reporting less than 60 per cent) for tax reasons, decreased from 30 per cent in 1999, to 23 per cent in 2002, and 12 per cent in 2005. However, even in 2005 only two-thirds of all firms show 'complete reporting' (90–100 per cent of total sales).

There are also considerable differences in tax compliance depending on firm size. The 2005 survey results reveal that truthful sales reporting seems to be more common among large enterprises (75 per cent of all firms report full compliance) than small firms, where it seems that only some 60 per cent report fully on sales. One implication of this is that it may make it possible for countries to run countercyclical fiscal policies[15] and to continue to invest during times of economic downturn. Experience would suggest that, as is the case with R&D spending, procyclical spending on education should be avoided. For the non-resource-rich transition countries in particular, the best form of countercyclical spending would be debt finance. There is undoubtedly scope to change the composition of spending, with a greater share being allocated to education on the grounds that it enhances growth.

The situation facing the resource-rich countries of the CIS+M is different and there is the potential to finance growth-enhancing reforms, even at 2010 levels of tax revenue and enforcement. For example, there could be stricter 'earmarking' of revenue for specific policy areas, particularly if the resources have been accumulated in stabilization funds (the case of Azerbaijan, Kazakhstan and Russia). This accumulated revenue has been used for general budget support rather than specific purposes, but the case could be made for it to be targeted to particular policy areas, such as education, that would produce positive consequences for long-term growth. We have shown that, as resource prices increase, the actual share of spending on education has tended to fall in the resource-rich countries – an undesirable outcome given the already relatively low levels of expenditure on education. Better targeting of resources within a medium-term financing framework could help to rectify this.

While governments have a major role in ensuring quality and access to formal education, the role of the private sector is crucial for ensuring a match between formal skills and labour market requirements. This can be achieved through internships, informal on-the-job training and funding in the form of scholarships for university courses. Such firm- and/or industry-specific investment would promote the skills that would increase labour productivity and labour mobility and, therefore, could produce direct spillovers that would benefit the whole economy.

However, to invest in enhancement of training and skills, firms need access to medium-term credit. There is widespread evidence that, despite changes in bank ownership and growth in non-bank finance, firms in the transition regions find it difficult to access formal credit markets. For firms that are relatively far from the technological frontier, better availability of bank credit is crucial. For firms close to the frontier, innovation will be riskier than imitation and these firms typically will have fewer capital resources to use as collateral in case of repayment default. Equity financing can compensate the financier for this added risk, typically by letting them get a higher share of returns.

3.8   CONCLUSIONS

This chapter has reviewed the growth experiences of transition countries. We argue that general and non-sector-specific government intervention can substantially increase the long-term growth prospects for these countries. While specific sectors might also benefit from specific policy measures, this chapter focused on overarching growth-enhancing policies,[16] and two areas in particular where policy can be effective: competition and education quality. For the transition countries to achieve and maintain higher growth rates in the long run, it will be necessary to encourage competition by continuing to remove entry and trade barriers and by strengthening and, in some cases, setting up competition agencies. This applies particularly to the CIS+M, and the resource-rich countries. Also, the transition countries as a group need to invest in the quality of primary and secondary education, which in turn implies greater investment in tertiary training, especially undergraduate education, in order to improve teacher quality and evaluation and monitoring of the overall education system. Somewhat paradoxically, it is the resource-rich CIS+M group of countries that have the lowest levels of investment in education and the greatest problems with quality of the education services provided. We discussed the scope for macroeconomic policies that would boost spending on these key areas. For education, in particular, intervention by the private sector to ensure appropriate skills training will benefit from greater financial intermediation and easier access to finance.

NOTES

1. Principally Singapore, Hong Kong, Taipei China and South Korea; see Hausmann et al. (2005).
2. Spence and the Commission on Growth and Development (2008).
3. Patent applications are filed with national patent offices for exclusive rights to an invention (product or process) that provides a new way of doing something or offers a new technical solution to a problem. Increasingly, patents applied for in transition countries are also registered internationally with the European Patent Office, the US Patent and Trademark Office and the World Intellectual Property Office.
4. The technological frontier is the current international limit of technological capabilities in a specific sector. In most empirical analysis, current US technology is used to proxy the technological frontier.
5. The higher the remaining profit margins of a given firm, the higher that firm's market power and, thus, the lower the overall level of competition in that particular market.
6. It can be argued that increased competition should discourage innovation and growth since it reduces the rewards that accrue to successful innovators. However, this effect may be dominated by increased competition, which may encourage firms to innovate in order to escape competition.
7. Benhabib and Spiegel (1994), drawing on seminal work by Nelson and Phelps (1966).
8. Equivalent to 1 standard deviation in the PISA results for OECD countries.
9. The long-term effects are based on simulations.
10. What holds for between countries holds also for between regions within a country. For example it has been shown that, in a US state, an additional US$1000/person spending on research education raises productivity growth by 0.27 per cent if the state is at the frontier, but by only 0.09 per cent if the state is far from the frontier; see Aghion, Boustan et al. (2005) and Vandenbussche et al. (2006).

11. Gross enrolment ratio is the ratio of total enrolment, regardless of age, to the population of the age group that officially corresponds to the level of education shown.
12. The PISA (Programme for International Student Assessment) study was carried out by the OECD in 2000, 2003 and 2006. It is one of the few sources of international comparative data on education across regions (including a number of transition countries), measuring educational quality by testing the mathematics, science and readings skills of a sample of 15-year-old students. The PISA surveys make a particular effort to assess students' skills in application and synthesis of concepts – the generic skills that are most relevant to the needs of the global economy. See Mertaugh and Hanushek (2005).
13. See EBRD *Transition Report* (2007). The survey found also that around half of respondents put additional government investment in education among their top priorities.
14. See BEEPS data.
15. Countercyclical in this context means continuing investments independent of the economic cycle, that is committing a similar amount of resources to education in times of both boom and downturn. Procyclical investments are aligned with the overall economic cycle, increasing in boom periods and decreasing in economic downturns.
16. For a sector-specific analysis of innovation policy see, for example, Boeheim et al. (2009).

REFERENCES

Aghion, P. and U. Akcigit (2008), 'Appropriate finance and growth', mimeo.

Aghion, P. and P. Howitt (2006), 'Joseph Schumpeter Lecture – Appropriate growth policy: a unifying framework', *Journal of the European Economic Association*, **4**(2–3), 269–314.

Aghion, P., Y. Algan and P. Cahue (2008), 'Can policy interact with culture? Minimum wage and the quality of labour relations', *NBER Working Paper No. 14327*, Washington, DC: National Bureau for Economic Research.

Aghion, P., W. Carlin and M. Schaffer (2002), 'Competition, innovation and growth in transition: exploring the interactions between policies', *William Davidson Institute Working Papers Series No. 501*, Ann Arbor, MI: William Davidson Institute, University of Michigan.

Aghion, P., T. Fally and S. Scarpetta (2007), 'Credit constraints as a barrier to the entry and post-entry growth of firms', *Economic Policy*, **22**(52): 731–79.

Aghion, P., L. Boustan, C. Hoxby and J. Vandenbussche (2005), 'Exploiting states' mistakes to identify the causal impact of higher education on growth', Working Paper, Harvard University, Department of Economics, Cambridge, MA, available at: http://www.economics.harvard.edu/faculty/aghion/files/Exploiting_States_Mistakes.pdf, accessed 7 August 2010.

Aghion, P., N. Bloom, R. Blundell, R. Griffith and P. Howitt (2005), 'Competition and innovation: an inverted-U relationship', *Quarterly Journal of Economics*, **120**(2), 701–28.

Ammermueller, A., H. Heijke and L. Woessmann (2005), 'Schooling quality in Eastern Europe: educational production during transition', *Economics of Education Review*, **24**, 579–99.

Barro, R.J. and X. Sala-i-Martin (1995), *Economic Growth*, New York: McGraw-Hill.

Benhabib, J. and M. Spiegel (1994), 'The role of human capital in economic development: evidence from aggregate cross-country data', *Journal of Monetary Economics*, **34**(2), 143–74.

Blundell, R., R. Griffith and J. Van Reenen (1999), 'Market share, market value and innovation in a panel of British manufacturing firms', *Review of Economic Studies*, **66**, 529–54.

Boeheim, M., A. Reinstaller and F. Unterlass (2009), 'Innovation policy from a sector specific perspective: implications for the New Member States', paper presented at the Innovation for Competitiveness INCOM Prague workshop, 22–23 January.

Carlin, W., M. Schaffer and P. Seabright (2004), 'A minimum of rivalry: evidence from transition economies on the importance of competition for innovation and growth', *Contributions to Economic Analysis and Policy*, **1**, 1284–327.

EBRD - World Bank (various years), 'Business Environment and Enterprise Performance Survey (BEEPS)', available at http://www.ebrd.com/pages/research/analysis/surveys/beeps.shtml.

European Bank for Reconstruction and Development (EBRD) (2006), *Transition Report 2006: Finance in Transition*, London: EBRD.

European Bank for Reconstruction and Development (EBRD) (2007), *Transition Report 2007: People in Transition*, London: EBRD.

European Bank for Reconstruction and Development (EBRD) (2008), *Transition Report 2008: Growth in Transition*, London: EBRD.

Hanushek, E.A. and D.D. Kimko (2000), 'Schooling, labor force quality, and the growth of nations', *American Economic Review*, **90**(5), 1184–208.

Hanushek, E.A. and L. Woessmann (2008), 'The role of cognitive skills in economic development', *Journal of Economic Literature*, **46**(3), 607–68.

Hausmann, R., D. Rodrik and A. Velasco (2005), 'Growth diagnostics', John F. Kennedy School of Government, Harvard University, Cambridge, MA, available at http://ksghome.harvard.edu/~drodrik/barcelonasep20.pdf, accessed 7 August 2010.

Levine, R. (2004), 'Finance and growth: theory and evidence', *NBER Working Paper Series No. 10766*, Cambridge, MA: National Bureau of Economic Research.

Mertaugh, M. and E. Hanushek (2005), 'Education and training', in N. Barr (ed.), *Labor Markets and Social Policy in Central and Eastern Europe: The Accession and Beyond*, Washington, DC: World Bank, Chapter 7.

Nelson, R. and E. Phelps (1966), 'Investment in humans, technological diffusion and economic growth', *American Economic Review*, **56**(1–2), 69–75.

Nickell, S.J. (1996), 'Competition and corporate performance', *Journal of Political Economy*, **104**(4), 724–46.

OECD (2007), *PISA 2006 Science Competencies for Tomorrow's World*, Paris: OECD.

Spence, M. and the Commission on Growth and Development (2008), *The Growth Report: Strategies for Sustained Growth and Inclusive Development*, International Bank for Reconstruction and Development/World Bank, on behalf of the Commission on Growth and Development, available at http://www.growthcommission.org, accessed 23 April 2010.

Vandenbussche, J., P. Aghion and C. Meghir (2006), 'Growth, distance to frontier and composition of human capital', *Journal of Economic Growth*, **11**(2), 97–127.

World Bank (2000), 'Educational strategy in ECA – Hidden challenges to education systems in transition economies', Human Development Sector, Washington, DC.

World Bank (2005), 'Education sector strategy update: achieving education for all, broadening our perspective, maximizing our effectiveness', presented to the Board of Directors on 17 November, World Bank, Washington, DC.

World Bank (2006), *Expanding Opportunities and Building Competencies for Young People: A New Agenda for Secondary Education*, Washington, DC: World Bank.

World Bank (2008), 'Doing business', available at http://www.doingbusiness.org/, accessed 23 April 2010.

4. Sectoral innovation modes and level of economic development: implications for innovation policy in the new member states

Andreas Reinstaller and Fabian Unterlass

4.1 INTRODUCTION

OECD data show that most European countries, and the European Union (EU) as an economic area, are lagging behind the US by some 30 per cent of per capita gross domestic product (GDP) and labour productivity (OECD, 2007). Composite indicators, such as the World Economic Forum's Global Competitiveness Index or the European Innovation Scoreboard (EIS), suggest that this lower economic performance is related to institutional and structural deficits that have a negative effect on the competitiveness of European countries. It is widely accepted that Europe's economy has lost ground to the US because of lack of competition and entrepreneurship, poor investment in higher education, insufficient credit market funding for growing enterprises and procyclical fiscal policies, all of which are detrimental to growth (Sapir et al., 2003; Bartelsman and van Ark, 2004; Aghion and Howitt, 2006; EC, 2008). As a result, the US and Japan are attracting more international research and development (R&D) investment than the EU. The US is also more successful at attracting top researchers and highly skilled staff (EC, 2005). The conditions that nurture the generation of new knowledge, innovative industries and, thus, innovation-based growth, are considered more favourable in the US and Japan than in Europe.

Institutional reform and industrial restructuring for greater innovation and growth are the principal goals of the Lisbon Growth and Jobs Strategy promoted by the European Council and the European Commission (EC). This strategy calls for national reform plans to address all areas of policy pertinent to the competitiveness of EU member states. EU member states have the flexibility to implement whatever measures they consider are most appropriate to advance their innovation potential. However, despite the breadth of the Lisbon Agenda and the flexibility it encompasses, most member states are focusing on the quantitative goal of increasing national average R&D investment in Europe to 3 per cent of EU GDP, the so-called Barcelona target (EC, 2002).

Academics are increasingly sceptical about this policy focus. Van Pottelsberghe (2008), for instance, shows that countries with the highest gap in 2004 between national R&D intensity and the 3 per cent EU spending target, have pledged to increase their investment so that, rather than 3 per cent being the average for all member states, all countries should achieve this goal. In particular, the new member states (NMS), where R&D investment was lagging the most against the average for Europe, have committed to a substantially increased share of R&D spending on GDP. However, differences in industry specialization or technological capability may require member states to develop more differentiated policy, which goes beyond a simple input goal. What is important, therefore, is what would be the most appropriate science, technology and innovation (STI) policy for both individual EU member states and the EU as a whole. This chapter aims to provide a better understanding of this issue.

In section 4.2 we discuss the limitations of the 3 per cent Barcelona goal. We show that a more differentiated perspective is needed to develop appropriate approaches to STI policy across EU member states. We propose some country and industry classifications, which – at least from an analytical point of view – are helpful to develop this differentiated perspective. In section 4.3, we present an empirical analysis of the differences in innovation behaviour based on these country–industry classifications. In section 4.4, based on these results, we draw some conclusions about how adequate provision for national specialization profiles and sectoral innovation modes may help in the design of policy mixes appropriate for fostering innovation, competitiveness and growth.

4.2 THE 3 PER CENT GOAL IN PERSPECTIVE: NATIONAL R&D PERFORMANCE AND INDUSTRY STRUCTURE

National R&D performance and national industry structure are closely linked. If high R&D-intensive sectors contribute unevenly to GDP across countries, then, due to this structural effect, aggregate R&D intensity might be higher in some countries than others, despite R&D investment for the industries in question being the same across countries. This would imply that in order to increase national R&D intensity, it may be necessary to induce structural change towards more technology-intensive sectors in some countries, and to increase the R&D intensity of all industries in others, but without the need to support structural change. International comparisons of R&D intensities that do not take account of individual country specializations may lead to erroneous policy conclusions. To highlight this, we decompose manufacturing sector R&D intensity in the EU member states into structural and country-specific components. The method of adjustment adopted in this chapter follows Sandven and Smith (1998); we calculate it as follows:

$$RD_i = \sum_{j=1}^{n} \overline{RD_j} * w_{i,j} + \sum_{j=1}^{n} (RD_{i,j} - \overline{RD_j})\, w_{i,j} \qquad (4.1)$$

where RD_i is manufacturing sector R&D intensity in country i, $RD_{i,j}$ is the value of the measure for sector j in country i, $\overline{RD_j} \equiv 1/m\Sigma_{i=1}^{m} RD_{i,j}$ is the average of the measures of sector j across all m countries in the sample and $w_{i,j}$ is the fraction of gross output or value added produced by sector j as a share of total value added or gross output in country i.[1] The first term on the right-hand side of equation (4.1) gives the structural component, which represents the R&D intensity we would expect from the country's industrial structure. The country-specific component represents how much each industry's R&D intensity deviates from the expected value. The country-specific component, therefore, captures the differences in innovation behaviour of the sectors in a given country from the innovation behaviour of a 'representative' sector, given by the average across industries. It reflects country-specific effects that influence sector performance.

Table 4.1, which is based on countries for which data are available, shows that the industry structure explains between 11 per cent and nearly 80 per cent of aggregate manufacturing sector R&D intensity in the EU (column 4). It shows also that there is considerable divergence in the NMS between expected and observed manufacturing sector R&D intensity (column 3, derived from column 2 minus column 1). This indicates that there are considerable country-specific effects and that R&D spending in the NMS is below the levels of these countries' specialization profiles. It shows that national specialization has an important impact on national R&D intensity (column 4). However, there is a considerable variation in the size of the country effects across member states.

Table 4.1 shows that manufacturing sector R&D intensity in the NMS and in Belgium (BE) and the Southern European countries, Italy (IT), Portugal (PT), Spain (ES) and Greece (GR), is lower than their industry structures would indicate,[2] while Sweden (SE), Germany (DE) and France (FR) show higher than expected values.[3] It suggests also that the industrial

Table 4.1   Structural decomposition of intramural R&D spending

Country	R&D intensity %	Expected R&D intensity %	Country effect %	Sector effect %
EU15 countries				
Belgium	1.299	1.783	−0.48	72.85
France	2.462	1.850	0.61	75.14
Germany	2.456	1.942	0.51	79.07
Greece	0.714	1.19	−0.48	59.63
Italy	0.814	1.56	−0.75	51.98
Portugal	0.311	1.125	−0.81	27.72
Spain	0.622	1.555	−0.93	40.03
Sweden	3.886	2.030	1.86	52.24
New member states				
Cyprus	0.215	1.025	−0.81	21.00
Czech Republic	0.566	1.631	−1.06	34.71
Estonia	0.435	1.329	−0.89	32.76
Hungary	0.288	1.937	−1.65	14.89
Lithuania	0.481	1.082	−0.6	44.42
Poland	0.168	1.385	−1.22	12.15
Romania	0.277	1.272	−0.99	21.78
Slovak Republic	0.168	1.574	−1.41	10.71
Non-EU countries				
Norway	1.34981	1.75926	−0.41	76.73

Notes: Column 1 Manufacturing sector R&D intensity; Column 2 Expected R&D intensity (sector effect) based on national industrial structure; Column 3 Country effect (CE); Column 4 Share of national R&D intensity explained by national sector structure, calculated as 1-abs(CE)/(abs(CE)+SE).

Source: Adapted from Eurostat New Chronos, CIS-4 and WIFO.

structures of the Czech Republic and Slovakia show specialization in medium R&D intensity sectors, with expected values slightly lower than those for Germany or France. The results show also that Hungary is specialized in high R&D intensity industries, while Portugal and Greece and a number of NMS, such as Lithuania, Latvia, Romania and Poland, are specialized in low technological intensity sectors.

Our results clearly support the views of those who are sceptical about the 3 per cent goal. In the absence of clear information on national technological specialization and technological capabilities, this level of R&D

investment may not provide the best incentive for industry innovation, or foster industrial restructuring aimed at technology-intensive industries. At the same time, in countries specialized in technology-intensive industries 3 per cent may not be a sufficiently ambitious goal, and for countries specialized in low-technology sectors it may promote industrial restructuring, but only if increased R&D spending is supported by measures addressing complementary aspects of their national systems of innovation (NSI), such as the availability of high-skilled human resources and university research, which increase absorptive capacity and research productivity.

There is another argument against this 3 per cent target, which is that there seems to be no one industrial structure that is conducive to growth and the creation of more and better jobs. In some industries, output growth is driven by productivity improvements; in others, it depends on demand growth (see Harberger, 1998; Smith, 1999; Hölzl and Reinstaller, 2007). Some industries are innovative, but conduct very little R&D since their innovation activities rely on inventions developed elsewhere; others rely heavily on in-house research. Also, whether an alleged high-tech industry is really technology-intensive depends on the position of its firms in the international supply chain. The outputs of a technology-intensive industry may be high-tech, but many of the production steps and intermediate products in the industry may not be. It would be misleading, therefore, for policy to focus exclusively on encouraging R&D investment and supporting restructuring towards R&D-intensive industries as a way to achieve growth and employment. Ultimately, it is the successful transformation of different production factors into innovative outputs that determines the competitiveness of firms. R&D is an important, but not the only input factor in the innovation process, and technology-intensive firms and industries are important, but not the only drivers of growth. In order to develop more differentiated and better targeted science, technology and innovation (STI) policies it is necessary to have a more detailed understanding of the factors that drive innovation across countries and industries.

4.3 DIFFERENCES IN INNOVATION BEHAVIOUR ACROSS COUNTRIES AND INDUSTRIES, AND DEVELOPMENT OF 'APPROPRIATE' STI POLICIES

The heterogeneity of countries' economic and social characteristics makes analysis of a large economic area, such as the EU, difficult, and renders statistical inference and comparative study across the units of observation less than straightforward. All of this, in turn, makes it difficult to draw meaningful policy conclusions. In addition, the smaller the unit of analysis, the greater will be the diversity. The effect is averaged across country data and is most obvious at the level of individual households and firms. As innovation occurs essentially at firm level, we need to find a way to deal with the multiple sources of heterogeneity. One solution is to use country and industry classifications that allow meaningful groups to be built, based on a few salient types, to analyse innovation behaviour within these sub-groups.

4.3.1   Classifying Countries according to Technological Profile

Advances in the theory of economic growth (Acemoglu et al., 2002) show that the factors driving it vary in function of the distance from the world technological frontier. Technological and market opportunities depend on the state of development. Countries that are far from the technological frontier typically face lower factor costs and are able to exploit cost advantages, while for countries close to the frontier factor costs are generally high and comparative advantage is based on the capability to innovate. Similarly, Lazonick (2002) claims that the way entrepreneurial activity contributes to the process of economic development cannot be analysed without an understanding of the link between the state of a country's economic development and its innovation activity. The first step in the analysis presented in this chapter, therefore, is to find a sensible way to classify countries according to technological potential.

Figure 4.1 depicts the results of an input–output analysis carried out by Knell (2008) and provides quantitative evidence on embodied flows of technology between industries and countries via intermediate inputs and capital goods, and the intrinsic technological intensity of industries and countries. It provides evidence on the underlying reasons for the patterns in Table 4.2. Knell's analysis distinguishes between the real R&D intensity of a national industry and R&D embodied in domestic and foreign intermediate inputs and capital goods. It provides an accurate picture of the technological profiles of countries in terms of domestic technology intensity and reliance on technology flows from abroad.[4] The observations are ordered in decreasing domestic R&D components (own R&D and R&D embodied in domestic intermediate inputs and capital goods).

Figure 4.1 shows that in aggregate, some countries invest heavily in both internal R&D and R&D-intensive intermediate inputs and capital goods used in production. This is the case for countries that generally are considered to be innovation leaders, such as the USA, Sweden and Finland. For countries generally categorized as medium-tech, such as

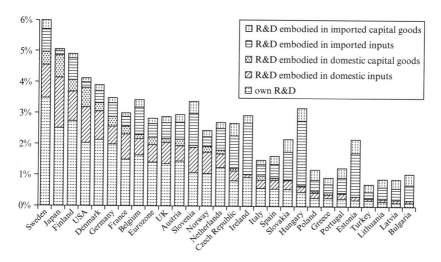

Source: Knell (2008) based on OECD ANBERD data, EUROSTAT Input–Output tables. NIFU-STEP calculations.

Figure 4.1   *Percentage share of total direct and indirect R&D intensity in the economy*

Austria, Belgium, France and the Netherlands, shares of both own R&D and R&D embodied in domestic inputs are smaller than for the innovation leaders. However, in these latter countries the direct and indirect components of domestic R&D are more significant than R&D embodied in foreign inputs. The NMS countries fall into two groups. The first group, which includes Hungary, Estonia, Slovenia, the Czech Republic, Slovakia and – perhaps surprisingly – Ireland, has relatively low own R&D intensity, and heavy reliance on R&D embodied in imported inputs; in other words, these countries rely on technology-intensive capital goods and intermediate inputs. In the second group of countries – Poland, Lithuania, Latvia, and Bulgaria, Greece and Portugal – the levels of all components of R&D investment are very low, and there is heavy reliance on imported technology.

It is possible statistically to construct a country classification from the data presented in Figure 4.1, to distinguish between predominantly technology users and predominantly technology producers. In our definition, a technology user country relies on technology transfer, while a technology producer invests heavily in own R&D. We use these country groupings to analyse differences in innovation behaviour at firm level across groups. By summing domestic direct and indirect R&D components we

Table 4.2   Country groups based on direct and indirect R&D intensity

	Country group	Countries
1	High direct and indirect R&D intensity	JAP, USA, SE, FI
2	Average direct and indirect R&D intensity	DK, DE, FR, UK, AT, BE NL, NO
3	Technology users with technology-intensive industries	HU, IE, EE, CZ, SK, SI
4	High-income countries with below-average direct and indirect R&D intensity	IT, ES, PT, GR
5	Low-income countries with below-average direct and indirect R&D intensity	PL, BG, LT, LV, TR

Source: Own analysis; data for cluster analysis adapted from Knell (2008); GDP per capita data are from Eurostat.

obtain a measure for the overall technological intensity of a country's production; by summing the components of foreign indirect R&D we obtain a measure of a country's dependence on foreign technology. We use these two measures, and GDP per capita at purchasing power parities to perform a cluster analysis.

The cluster analysis uses a hierarchical cluster method with all data standardized to the EU average for each variable. We use the normal and the squared Euclidian distances as our dissimilarity measure. This cluster algorithm is the most robust average linkage method, and we also use non-hierarchical methods to check robustness. Summary results of the cluster analysis are presented in Table 4.2. All cluster results are robust to identifying Group 1 (USA, JP, FI, SE), Group 2 (NO, FR, DE, DK, AT, UK, NL, BE) and Group 3 (EE, IE, SK, CZ, SI, HU). The non-hierarchical methods vary – sometimes categorizing Denmark (DK) as Group 1. All methods and all data combinations (with per capita GDP) consistently categorize Ireland in Group 3. The results for Groups 4 and 5 are generally less robust. Most methods allocate all the countries concerned to one group; hierarchical methods using per capita GDP differentiate the two groups. We believe this latter classification to be the most plausible and most meaningful: Group 4 includes Greece, Portugal, Spain and Italy, and Group 5 includes Poland, Bulgaria, Latvia, Lithuania and Turkey.

On the basis of all this evidence combined, we can conclude that industry structure plays an important role in explaining national R&D intensity. However, there are considerable country-specific effects. The results of an input–output analysis suggest that these effects are related to the extent that countries rely on domestic research investment or on embodied technology transferred from abroad. Countries that rely heavily on embodied technology flows from abroad and invest little in their own R&D show negative country effects, as reported in Table 4.1.

The picture is reversed for countries with high levels of domestic R&D. Hence, while Hungary and the Czech Republic, for example, are specialized in medium- to high-tech industries, the input–output results reveal that the production of industries in these countries uses technology from abroad, an indication that these countries work in less technology intense value chain segments. The NMS in the second group, however, have low to medium technological intensity industrial structures, and this is evident in their R&D investment patterns. However, levels are still considerably lower than for the EU15 countries, which have similar specialization profiles. This, in turn, is likely to be related to technological capabilities and the general economic development conditions in these countries.

4.3.2 Industry Classifications based on Innovation Profiles

Figure 4.2 depicts the firm-level determinants of innovation as a nested set of factors. Firms work within specific national contexts, whose related framework conditions affect managerial decisions, availability and cost of resources, costs of knowledge production and acquisition, and so forth. Various studies underscore the importance of competition, human capital formation, credit constraints, intellectual property rights (IPR) protection and the social conditions for the growth and performance of enterprises (Acemoglu et al., 2002; Aghion and Howitt, 2006; Aghion et al., 2005, 2007, 2009; Lazonick, 2002). There is also a large body of literature that analyses how the design and specifics of national innovation systems affect innovation behaviour at industry and firm levels (for example Nelson 1994; Nelson and Sampat, 2001).

Moving from national to industry level, Figure 4.2 shows that a different set of factors affects the innovation behaviour of firms. The Schumpeterian literature provides robust evidence that differences in the innovation behaviour of firms across industries are due to differences in the cumulativeness and appropriability of knowledge, and the opportunity conditions in an industry. These factors define the particular knowledge and learning environments in which firms operate and, therefore, affect the behaviour of the firms in each industry (see Malerba and Orsenigo, 1995).

Cumulativeness of knowledge refers to the extent to which a firm's

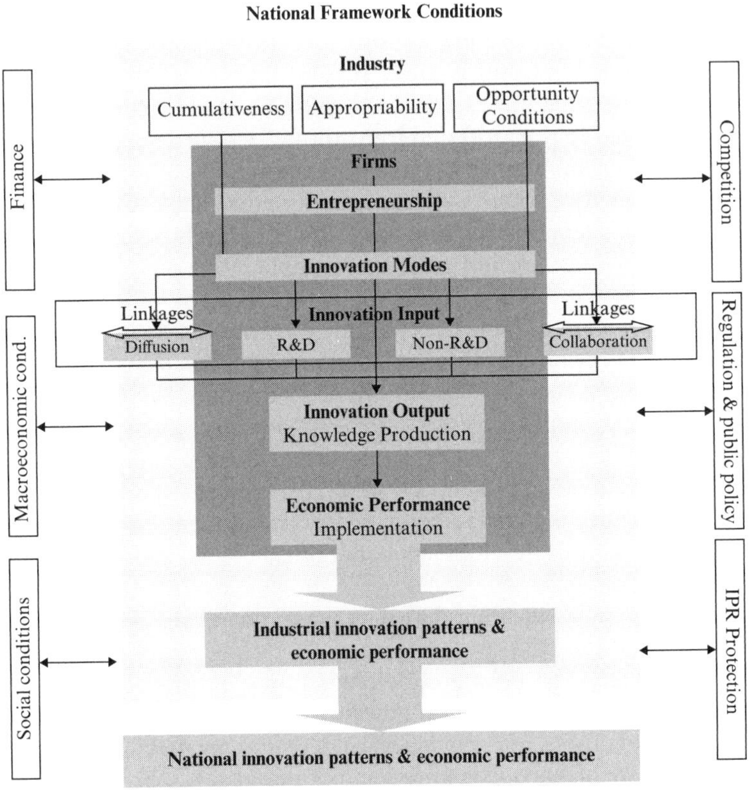

National Framework Conditions

Source: Adapted from Reinstaller and Unterlass (2008a: 26).

Figure 4.2   National framework conditions

ability to create new knowledge depends on the stock of knowledge accumulated in the past. Appropriability conditions refer to the possibilities
of firms to protect their innovations from imitation and, therefore, to the
possibility to extract profits from them. Both these aspects depend on a
number of factors, such as the complexity of the technology, the tacitness
and codifiability of the knowledge, and the marketing and research sunk
costs. Finally, opportunity conditions refer to the likelihood of producing
an innovation for a given sum of money; they depend on the technologies
used in the industry and on the characteristics of demand.

Studies have established that industry characteristics, such as market
structure, average firm size, and patterns of innovation expenditure, are
closely related to these factors (see Levine et al., 1987; Klevorick et al.,

1995). For instance, there is robust evidence that in industries where opportunities are high and appropriability and knowledge cumulativeness are low, innovations are usually introduced by start-up firms. These sectors are characterized by a process of 'creative destruction', in which many firms enter the industry and many others leave it. Consequently, market concentration is low and firms are small sized: the machinery industry is an example here. If, on the other hand, the knowledge base is proprietary and cumulative, then sectors typically follow a pattern of 'creative accumulation', in which large firms dominate and industry concentration is high (see Breschi et al., 2000), exemplified by the automotive industry. These and other commonalities in the innovation behaviour and performance of firms can be identified as sectoral innovation modes (for example Robson et al., 1988; Malerba, 2002, 2004).

Within these sector-specific boundary conditions there is considerable heterogeneity of innovation behaviour at firm level, which is dependent upon the ways that firms try to exploit promising conditions. Firms can seek actively to exploit these conditions through their own process or product innovations, or to explore opportunities through imitation, technology adoption or opportunities not related to technology. Following Peneder (2007), the first type of firm behaviour can be described as 'creative entrepreneurship', and refers to firms that seek to differentiate from their competitors by creating new technologies or products. These types of entrepreneurs are innovation leaders. The second type of firm behaviour can be called 'adaptive entrepreneurship'. This label describes firms that try to close in on the innovation leaders or that seek to diversify through activities other than technological innovation. Entrepreneurship, therefore, is a firm-specific characteristic.

Peneder (2007) constructs an innovation classification based on Community Innovation Survey (CIS) micro data for 21 countries. He classifies firms on the basis of entrepreneurship type and technological regimes (Box 4.1).

As patterns of entrepreneurial behaviour find expression in the exploration strategies pursued by firms, other things being equal, richer technological opportunities make research activities potentially more profitable (see for example Pakes and Schankerman, 1984; Nelson, 1994; Nelson and Wolff, 1997). Creative entrepreneurs, therefore, are more likely to be observed in industries with high technological opportunities, where the creation of new technologies or products is R&D driven. In industries where technological opportunities are less abundant, adaptive entrepreneurs are more frequent since firms are more likely to pursue other opportunities and access critical knowledge through means other than R&D. Such firms are more likely to rely on technologies and knowledge

---

BOX 4.1 INDUSTRY CLASSIFICATION BASED ON APPROPRIABILITY, OPPORTUNITY, CUMULATIVENESS AND ENTREPRENEURSHIP

Entrepreneurship: The firm classification distinguishes between creative and adaptive entrepreneurship. Creative entrepreneurs are characterized by firm-specific innovations and can be further separated into firms producing: (i) their own process innovations; (ii) their own, new-to-the-market product innovations; or (iii) both. All other firms are characterized as adaptive entrepreneurs. Among these Peneder distinguishes a fourth group of technology adopters which create product innovations that are new to the firm, but not to the market, or process innovations mainly in cooperation with other enterprises or institutions. Finally, he identifies a fifth, residual group of adaptive entrepreneurs that pursue opportunities other than technological innovation.

'Technological regimes' are characterized in terms of opportunity, appropriability and cumulativeness conditions, whose combination defines the particular knowledge and learning environments within which the firm operates.

Opportunity conditions: The classification distinguishes four firm conditions according to the perceived technological opportunities demonstrated by the firm's innovation activity: (i) no opportunities – the firm neither performs intramural R&D nor purchases external innovations; (ii) acquisition – the firm innovates only by purchasing external R&D, machinery, or rights (patents, trademarks, etc.); (iii) intramural R&D – the firm undertakes its own R&D, but the ratio of innovation expenditure to total turnover is less than 5 per cent; and (iv) high R&D – the firm performs intramural R&D and its share of innovation expenditures in total turnover is more than 5 per cent.

Appropriability conditions: (i) strategic – for firms relying exclusively on secrecy, complexity of design, or lead-time advantages to protect their innovations; (ii) formal (other than patents) – firms that use the registration of design patterns, trademarks or copyright; (iii) patenting (either as well as or without strategic or other formal methods of protection); (iv) full arsenal – firms make use of all of the above three means of protection; (v) none – firms employ none of these tools.

Degree of knowledge cumulativeness: CIS data do not provide direct measures of cumulativeness. Peneder (2007) combines two aspects of the CIS data. First he differentiates according to the relative importance of internal vs external sources of information. Second, he applies contrasting identification rules depending on whether the firm seems to be a technology leader or a technology follower. Thus, firms within the 'creative response' classifications of entrepreneurship are characterized as operating within highly cumulative regimes if internal sources of knowledge are more or at least as important as external sources, and as operating in low cumulative regimes if the firm draws more on external than internal knowledge for its innovations. These identification rules are reversed for 'adaptive entrepreneurship' type firms.

not created inside the firm. Opportunity conditions and entrepreneurship, therefore, are closely related, but patterns of entrepreneurial behaviour vary systematically across sectors. They give rise to modes of innovation that persist over time.

Peneder's (2007) firm classification (Box 4.1) is used to construct a sector classification based on the frequency and combination of different firm types in each sector, presented in Appendix Table A4.1. Compared to other popular industry classifications, such as Pavitt's (1984) taxonomy, Peneder's has the advantage of being based on a very extensive set of firm-level data. The last column in Table A4.1 shows how these industry classes are exploited in the econometric analysis in this chapter.

4.3.3 Innovation Behaviour across Industry–Country Groups

Figure 4.2 illustrates that different entrepreneurial strategies related to knowledge sourcing and organizational innovations give rise to firm-specific modes of innovation. However, they are not unrelated to the properties of the technological regime and the general economic, technical and social conditions of the environment in which firms operate. We formulate some tentative hypotheses related to how opportunity conditions, knowledge sourcing and entrepreneurial strategies, as well as cumulativeness and appropriability, vary in function of the distance to the technological frontier, based on our reading of the literature.

Hypothesis 1a: Firms will exploit different opportunity conditions depending on the distance to the technological frontier. The importance of R&D and innovation investments more generally, will increase the closer firms are to the technological frontier. For firms that are far from the technological frontier, non-R&D-based and technology transfer-based methods will prevail.

Hypothesis 1b: For firms close to the technological frontier, the importance of universities as a source of methodological and basic science-related knowledge increases, and cooperation with universities positively and significantly affects their innovation output.

Hypothesis 1c: Since innovation investments are costly and uncertain, firms that spend more on this type of investment will be more active in acquiring IPR protection. Ceteris paribus, *in country–industry groups where firms invest more in innovation-related activities, IPR will be more important for innovation than in country–industry groups where this is not the case.*

These hypotheses are justified in so far as firms in less advanced countries will base their research mostly on appropriation, that is, they will conduct activities that enable them to adopt, imitate and modify innovations from more advanced firms, and will endeavour to exploit factor cost-based comparative advantages. Such activities are characteristic of firms operating in countries that are catching up, where growth is based on low factor costs and on scale economies rather than on innovation.

  The next group of hypotheses is related to the cumulativeness of knowledge.

Hypothesis 2a: Innovation is a cumulative process in so far as it involves the generation of high-quality or low-cost products based on a learning process which enables the recognition and exploitation of new opportunities. Technologically more advanced countries have higher accumulated stocks of knowledge and critical assets such as human capital. Firms located in these countries and operating in industries with high levels of cumulativeness will be more innovative than firms operating in countries where these stocks are not available.

Hypothesis 2b: An entrepreneur in an innovative enterprise has to engage in strategic control in order to allocate resources within an innovative investment strategy and to engage in the creation of incentives for the individuals participating in the firm's hierarchical and functional division of labour to apply their skills and efforts to the innovation process. New management techniques and organizational changes will be important in industries that are innovation-intensive. Given the cumulative processes involved, the instruments will vary across country groups depending on the distance from the technological frontier.

4.4   INNOVATION DETERMINANTS ACROSS COUNTRY AND INDUSTRY CLASSES

In this part of the chapter we explore differences in the drivers of firm innovation across the country and industry groups defined in the previous section, using firm-level data. The aim is to assess to what extent the determinants of firms' innovative output vary systematically across sectors and countries, given the technological profiles of the countries in which they are located and the technological profiles of the sectors in which they operate. The analysis models the share of firms' production that is new to the market, using a broad set of variables based on an innovation production function framework. Hence, the analysis provides an overview of how innovation behaviour varies across sectors and countries for the most innovative firms. This approach is justified since innovation implies the generation of new products that give better value for money, and because the firms we analyse are the innovation leaders that set the pace for the innovation followers in their country groups.

4.4.1   The Data

The findings in this chapter draw on the results of the Sectoral Innovation Watch Project, which was part of the EC Europe Innova initiative.[5] The project allowed access to CIS-3 micro-aggregated data for 15 countries and access to micro data for 21 countries, for the period 1998–2000. The results reported here refer to anonymized data for 15 EU countries and a total of more than 44 000 firms. CIS-3 provides data on the input and the output dimensions of innovation activities. In terms of innovation outputs, firms are asked to report the proportions of their turnover that are derived from products that are new to the firm, but not the market, and products that are new to both firm and market. We use the share of turnover from products new to the market as the principal output variable on which we regress a number of innovation input factors along the lines discussed in section 4.1. We use the sector and country classifications presented in Tables 4.2 and A4.1 to estimate the restricted models. Since the countries in the CIS sample do not exactly match the countries used to construct the country classification in Table 4.2, we need to allocate the observations for some countries to one or other of the groups. Based on the figures for gross R&D investment at country level, we allocated Romania to the country group that includes Bulgaria, Latvia and Lithuania, and Iceland to the group that includes Belgium and Germany. Next, because Group 1 in Table 4.2 includes a small number of countries, we combine Groups 1 and 2. Table 4.3 reports the sample averages for the dependent variable

Table 4.3   Average share of turnover from products new to the market (in %)

Pooled sample	Medium–high innovation intensity industries				Medium–low innovation intensity industries			
	Groups 1 & 2 (BE, DE, IS, NO)	Group 3 (ES, PT, GR)	Group 4 (SK, CZ, EE, HU)	Group 5 (BG, RO, LT, LV)	Groups 1 & 2 (BE, DE, IS, NO)	Group 3 (ES, PT, GR)	Group 4 (SK, CZ, EE, HU)	Group 5 (BG, RO, LT, LV)
~ 5% 6 435	~ 6% 1 305	~ 3.7% 1 717	~ 7.9% 1 231	~ 9.1% 704	~ 3.5% 487	~ 1.4% 399	~ 4.5% 775	~ 3.6% 493

Source: CIS-3 data

in our analysis, for the pooled sample, and for the sub-groups; the second data line reports the number of observations in each sub-group.

Table 4.3 shows that the average share of turnover from products new to the market varies considerably across country and industry groups. For medium to high innovation intensity industries, shares are above average; for medium–low to low innovation intensity industries, shares are below average. It is interesting that firms in high innovation intensity industries located in the NMS (Groups 4 and 5) have very high shares.

4.4.2   Econometric Strategy

We explore the hypotheses regarding the cross-country, cross-industry differences in innovation proposed in section 4.3.3, using an innovation production function approach which relates inputs to the innovation process to a measure of innovation output. As already discussed, the dependent variable in this analysis is the share of turnover from products new to the market. The independent variables capture the principal characteristics of the technological regimes and entrepreneurial exploration strategies, as well as firm-level, sector-level and country-level controls. The list of independent variables in the analysis and how they are constructed from original CIS data are presented in Appendix Table A4.2. We specify the following regression for the pooled sample of firm data and for eight country-industry groups:

$$InnoShare = \alpha + \sum_{n}^{4}\beta_{appr,n}d_{appr,n} + \beta_{cum,h}d_{cum,h} + \sum_{l}^{3}\beta_{opp,l}d_{opp,l}$$

$$+ \sum_{g}^{8}\beta_{sources,g}d_{sources,g} + \sum_{k}^{6}\beta_{strat,k}d_{strat,k} + \sum_{m}^{5}\beta_{fcontrols,p}d_{fcontrols,p}$$

$$+ \sum_{m}^{15}\beta_{country,m}d_{country,m} + \sum_{m}^{13}\beta_{NACE,m}d_{NACE,m} + \varepsilon. \qquad (4.2)$$

where $\beta_{i,j}$ are the coefficients for each of the dummy variables $d_{i,j}$ capturing the characteristics of technological regime, entrepreneurial exploration and the control variables specified in Appendix Table A4.1.

Our econometric strategy largely follows the methodology in Falk (2007). Due to the range of dependent variables being a share, we use a fractional logit model (Papke and Wooldridge, 1996) with robust standard errors, using a quasi-maximum likelihood estimator with heteroscedasticity robust asymptotic variance. Given that the CIS is not aimed at non-innovative firms, we need to correct for sample selection bias due to restricting the sample to innovation-active firms. To correct for this, we use a two-step Heckman model. We estimate a selection equation modelling the probability that a firm is innovative or not. From these estimations we construct the inverse Mill's ratio to correct for bias in the share equation (4.2). In our analysis, we use different specifications to model the selection equation. In the results reported here, we use total innovation expenditure, firm size class, firm age, and principal markets of activity. For this and other specifications, the correction factor turns out to be only marginally significant.

For each group of variables Wald tests for joint significance were carried out to verify whether just a few variables are significant separately, or whether all variables are capturing a specific aspect of entrepreneurship or technological regime. Table 4.4 summarizes the results of the regression analysis. For reasons of space, only parameter values and levels of significance are reported.

4.4.3   Results

4.4.3.1   Comparisons across country groups

The importance of high levels of innovation expenditure as a determinant of innovation output increases with proximity to the technological frontier   Levels of total innovation expenditure (including R&D) beyond 5 per cent of turnover are important across industry groups and especially for firms in countries that are close to the technological frontier. However, this variable is insignificant for firms in the most advanced countries, in high innovation intensity industries, but this is not to imply that this factor is unimportant. Finer analysis of the data shows that within this group, almost all firms invest more than 5 per cent of their turnover in innovation. Thus, this variable cannot discriminate between firms that perform well and those that perform less well. In other words, being close to the technological frontier in the most innovative industries means that high levels of innovation expenditure are a precondition for innovation activity.

Innovation intensive industries in less developed countries benefit from knowledge transfer and non-R&D-related knowledge sources   For industries with high to medium-high innovation intensity firms in the NMS, technology transfer is a highly important driver of innovation. For firms in the same industry group in the more advanced countries, this variable is not statistically significant. The case of low innovation intensity industries is interesting as results are insignificant, and significant and negative for firms in the less developed NMS. These results may be due to the fact that firms in this group often purchase second-hand machinery from abroad and exploit only comparative cost advantages. This view is supported by

Table 4.4 Regression results

Dependent variable: Share of turnover from sales of products new to the market	VARIABLES	pooled sample	Industries with medium to high innovation intensity				Industries with medium-low to low innovation intensity			
			Country groups 1 & 2 (BE, DE, IS, NO)	Country group 3 (ES, PT, GR)	Country group 4 (SK, CZ, EE, HU)	Country group 5 (RO, BG, LT, LV)	Country groups 1 & 2 (BE, DE, IS, NO)	Country group 3 (ES, PT, GR)	Country group 4 (SK, CZ, EE, HU)	Country group 5 (RO, BG, LT, LV)
appropriability	*p value Wald test*	*0.000*	*0.000*	*0.015*	*0.068*	*0.897*	*0.000*	*0.000*	*0.006*	*0.192*
	Strategic protection	0.209**	−0.0149	0.300**	0.293	0.0620	1.013***	0.0529	0.932***	−0.245
	Formal IPR protection	−0.271*	0.694*	0.121	0.392	0.217	−1.708***	−1.365***	0.690*	−0.509
	Patents	0.444***	0.748***	0.335***	0.759***	0.150	1.693***	0.0731	1.003**	0.374
	All methods	0.327***	0.757***	0.325**	0.472*	0.110	0.730*	−0.265	1.068***	−0.0588
opportunity	*p value Wald test*	*0.000*	*0.643*	*0.000*	*0.002*	*0.083*	*0.022*	*0.000*	*0.000*	*0.278*
	Tot. Innovation exp. share > 5% turnover	0.489***	−0.190	0.418***	0.381	0.170	0.584**	1.042***	1.180***	0.142
	Technology transfer	0.0349	−0.0785	0.0125	0.556**	0.327**	−0.221	0.142	0.132	−0.495***

Table 4.4 (continued)

Dependent variable: Share of turnover from sales of products new to the market	pooled sample	Industries with medium to high innovation intensity				Industries with medium-low to low innovation intensity			
VARIABLES		Country groups 1 & 2 (BE, DE, IS, NO)	Country group 3 (ES, PT, GR)	Country group 4 (SK, CZ, EE, HU)	Country group 5 (RO, BG, LT, LV)	Country groups 1 & 2 (BE, DE, IS, NO)	Country group 3 (ES, PT, GR)	Country group 4 (SK, CZ, EE, HU)	Country group 5 (RO, BG, LT, LV)
Non R&D related sources	0.178**	−0.286	0.239***	0.673**	0.179	−0.530	0.481***	0.443	0.231
cumulativeness Cumulativeness of knowledge	−0.0704	1.437***	0.0550	−0.335	−0.492***	0.574**	−0.987***	−1.322***	−0.287**
p value Wald test	*0.001*	*0.104*	*0.485*	*0.000*	*0.036*	*0.136*	*0.015*	*0.050*	*0.005*
sources of innovation Internal sources, high importance	0.162*	0.337*	−0.0373	0.843***	0.342**	−0.275	0.279	0.512**	0.127
Enterprise group as source, high importance	0.160	0.158	−0.0367	−0.456*	0.340	0.182	0.423*	0.226	−0.0759

Equipment suppliers, high importance	−0.0218	−0.282	−0.0871	0.151	−0.108	0.430	0.265	0.299	0.412**
Clients and customers, high importance	0.212**	0.0669	0.202	0.519**	0.0864	0.126	0.0448	0.0445	0.357**
Competitors, high importance	−0.164	0.0738	−0.216	−0.788***	−0.117	−0.527	−0.0476	−0.181	−0.203
Universities, high importance	0.348	0.724***	0.00700	−0.237	0.422*	0.000655	−0.718	−1.113**	−0.310
Government labs, high importance	−0.386*	−0.351	−0.146	−0.384	−1.080**	0.197	0.110	0.271	−0.844
Professional conferences, high importance	0.257*	0.0959	0.180	−0.00161	−0.0388	−0.923**	0.879***	0.0966	−0.208
Exhibitions, fairs, high importance	−0.0129	−0.288	−0.154	0.203	−0.0346	1.393***	−0.197	0.284	0.298
p value Wald test	*0.000*	*0.009*	*0.000*	*0.001*	*0.229*	*0.029*	*0.038*	*0.639*	*0.726*
New corporate strategy	0.143	0.256	0.167	0.140	0.382*	0.708**	−0.283*	0.0906	0.157

firm strategies

Table 4.4 (continued)

Dependent variable: Share of turnover from sales of products new to the market / VARI-ABLES	pooled sample	Industries with medium to high innovation intensity				Industries with medium-low to low innovation intensity			
		Country groups 1 & 2 (BE, DE, IS, NO)	Country group 3 (ES, PT, GR)	Country group 4 (SK, CZ, EE, HU)	Country group 5 (RO, BG, LT, LV)	Country groups 1 & 2 (BE, DE, IS, NO)	Country group 3 (ES, PT, GR)	Country group 4 (SK, CZ, EE, HU)	Country group 5 (RO, BG, LT, LV)
New management techniques	0.0274	0.289**	−0.149	−0.380*	−0.0995	−0.0918	0.124	0.324	0.0510
Change organization	0.0465	−0.0823	0.00683	0.617***	0.105	−0.335	0.225	0.118	0.196
New marketing strategy	0.0103	0.177	0.0596	0.514**	0.0201	0.102	−0.0239	0.0533	0.0177
Change in design of products	0.336***	0.115	0.454***	−0.0407	0.123	0.555**	0.456***	−0.0935	−0.163
p value Wald test	*0.000*	*0.000*	*0.000*	*0.083*	*0.000*	*0.000*	*0.002*	*0.094*	*0.001*
firm characteristics									
Cooperations	0.125	−0.0958	0.140	0.263	0.407***	0.131	0.282	0.156	0.112
Newly established	0.552***	1.179***	0.432**	0.483*	0.0822	0.570	0.305	−0.624	0.193

Increase of turnover due to merger >10%	0.268**	0.220	0.0921	0.445	0.177	0.621*	−0.662	0.765**	−0.133
Decrease of turnover due to merger >10%	0.0700	−0.375	−0.115	−0.116	0.0798	−0.155	0.439	0.246	0.916***
Internat. market	0.281**	0.375	0.167	0.814***	−0.150	0.534	0.148	0.546	0.0864
National market	0.124	0.0296	0.0490	0.496	0.0297	0.110	0.245	0.338	0.0728
Firm size quant. 75–100	−0.869***	−0.855***	−0.493***	−0.499*	−1.188***	−2.595***	−0.893***	−0.117	−0.667***
Firm size quant. 50–75	−0.492***	−0.616*	−0.306***	−0.491*	−0.595***	−1.196***	−0.353*	0.133	−0.696***
Firm size quant. 25–50	−0.211**	0.173	−0.105	−0.270	−0.367*	−0.940**	−0.246	−0.242	−0.274
Constant	−2.967***	−4.107***	−2.429***	−3.998***	−0.702***	−6.834***	−3.841***	−14.24***	−13.84***
Country dummies	YES	YES	YES	YES	YES	YES	YES	YES	YES
Sector dummies	YES	YES	YES	YES	YES	YES	YES	YES	YES
Sample selection correction	YES	YES	YES	YES	YES	YES	YES	YES	YES

Table 4.4 (continued)

Dependent variable: Share of turnover from sales of products new to the market	pooled sample	Industries with medium to high innovation intensity				Industries with medium-low to low innovation intensity			
VARI-ABLES		Country groups 1 & 2 (BE, DE, IS, NO)	Country group 3 (ES, PT, GR)	Country group 4 (SK, CZ, EE, HU)	Country group 5 (RO, BG, LT, LV)	Country groups 1 & 2 (BE, DE, IS, NO)	Country group 3 (ES, PT, GR)	Country group 4 (SK, CZ, EE, HU)	Country group 5 (RO, BG, LT, LV)
AIC	3.707	3.626	5.052	1.297	1.045	6.156	6.124	0.777	1.060
BIC	-122683.60	-13607.7	-31464.27	-15473.4	-6250.28	-2533.52	-7141.81	-13115.7	-4130.90
Pseudo LL	-29418.43	-4006.90	-13089.54	-1425.23	-487.38	-2393.76	-5259.29	-688.29	-339.22
deviance	30483.59	3248.84	12603.24	1716.60	532.51	2434.23	5431.72	786.51	347.70
R^2	0.256	0.361	0.166	0.274	0.202	0.612	0.243	0.314	0.143
Observations	6459	1305	1717	1231	704	487	398	775	493

Notes: Eurostat CIS-3 micro aggregated data. Fractional Logit model with robust standard errors. Regression weighted with the weights delivered with the dataset. The R^2 is calculated directly from the residual sum of squares. Sector groupings based on Peneder (2007). Standard errors and p-values omitted. Details available upon request from the authors. *** $p < 0.01$, ** $p < 0.05$, * $p < 0.1$; Wald tests on joint significance of variable group.

the fact that the only significant factors affecting innovation output for this group are contacts with equipment suppliers and customers. These and the previous results support Hypothesis 1a.

Cumulativeness in knowledge is highly important for firms in the advanced countries   The results for knowledge cumulativeness confirm that innovation is a highly cumulative and developmental process. High levels of knowledge cumulativeness significantly affect the innovation output of firms in countries that are close to the technological frontier. This indicates that innovation in these country groups requires firms to handle complex knowledge, which means that successful innovation requires experience. Interestingly, the effect diminishes and becomes negative for firms in less advanced countries. This holds true for both industry groups, although the intensity of the effect varies widely. Firms in industries with high knowledge cumulativeness operating in the NMS, seem to be generally less innovative than firms in the same industries in other country groups. This shows that firms in these countries have to overcome considerable barriers if they want to become competitive through innovation. This evidence supports Hypothesis 2a.

IPR protection is very important for firms in the most developed countries, but the effect diminishes for the less-developed country groups   Overall, our results indicate that IPR protection is highly relevant for innovation through its effect on the economic incentives for engaging in innovation. The share of turnover from market novelties is higher on average for firms that use IPR protection methods. The main method of protection is patenting. However, the effect differs across country groups. In the most developed group of the Northern and Central European countries (Groups 1 and 2) the effect is large; for the Southern member states (Group 3) and the more developed NMS (Group 4) the effect is small. In the less developed NMS, IPR protection plays no role. However, there are some differences between sector groups. In the case especially of the southern EU countries, IPR protection is irrelevant for low-tech sectors and is even negatively correlated with product innovation (formal IPR protection). The evidence supports Hypothesis 1c.

Small firms have higher shares of market novelties than large firms   The regression results show that firm size has a significant effect on the share of turnover from market novelties. With the exception of the more advanced NMS in low-tech sectors, this effect is consistent across country sector groupings. Larger firms produce significantly lower shares of market novelties. However, it may be that this result is due uniquely to a size effect. If a small firm launches a new product on the market the effect on the share of turnover from new products may be more accentuated than for equally innovative, but larger firms with well-established product portfolios.

The results are presented in Table 4.4.

4.4.3.2   Comparison of sector types

Newly established firms in sectors of high-innovation intensity are more innovative than older firms   Newly established firms have larger shares of market novelties in their turnover. However, in the group of technologically less developed NMS the coefficient is insignificant. This is especially interesting when considered in conjunction with the statement above. High entry rates of small and medium sized firms might be beneficial for innovation in sectors with high-innovation-intensive firms, but may play no role in low-innovation-intensive industries.

Mergers lead to higher shares of sales of innovative products in low-innovation intensity sectors. Organizational changes positively but unsystematically affect innovation output   In low-innovation-intensity sectors, new firms do not show significantly different correlations with innovation intensity than established firms. However, mergers are positively related to innovation performance. In the group of Northern and Central European countries (Groups 1 and 2) and the technologically more developed NMS countries, turnover growth due to mergers is correlated to increasing shares from market novelties. In the case of the less developed NMS, decreased turnover in the merged firms consistently increases the share of market novelties. This indicates that business consolidation may influence innovation performance positively. On the other hand, the results suggest also that the use of new management techniques, changes to corporate strategy, new marketing strategies, and changes to firm organization or product design, affect innovation performance positively. However, results are not systematic across country–industry groups and, hence, their interpretation is difficult. The evidence is inconclusive for Hypothesis 2b.

Internal sources are important for high-innovation intensity sectors, but almost irrelevant for low-innovation intensity sectors   Firms that are active in high-innovation intensity sectors generally show higher shares of turnover from market novelties if they rely on internal sources of innovation, that is, if innovations are generated within the firm. An exception is the Southern European countries (Group 3) where the results for low-innovation intensity sectors are different. For this group, it is only for firms in the more advanced NMS that this variable positively related to higher innovativeness; for the other three country groups the effect is insignificant, indicating that external sources of innovation are the most important for firms in these countries.

The use of knowledge sources differs widely across sector types and across country groups In the high-innovation intensity sectors, universities play a central role in knowledge sourcing in the Northern and Central European countries. Universities are important sources of advanced methodological knowledge and knowledge for complex problem solving. They support the capabilities of firms to develop and introduce market novelties in the most advanced countries. The results for the NMS are different in that universities play a different role in their national innovation systems in acting as technology transfer institutions. Hence, we observe a positive correlation for these countries. For low-innovation-intensive industries, fairs and exhibitions enable trade in knowledge and artefacts for innovation. Clients and customers are highly important for the more developed NMS, while firms that source their innovation knowledge from competitors are far less innovative. Both clients and suppliers are important sources of innovation for firms in the low-tech sectors in the group of less developed NMS with higher turnover from novelties. The results provide weak support for Hypothesis 1b.

The results seem generally to confirm the hypotheses proposed in section 4.3.3 and demonstrate the need for innovation policies that are commensurate with the state of economic development of each country.

4.5 CONCLUSIONS

The Lisbon Agenda, with its commitment to increasing the EU's R&D to GDP ratio to 3 per cent by 2010, is the linchpin of all the EU's efforts to foster innovation and growth. However, thus far, EU member states have failed to deliver on this goal. This can be attributed (partly) to the fact that 'EU manufacturing remains specialized in medium-tech sectors and has not taken advantage of the fast growth of certain high-tech sectors . . .' (EC, 2007: 5). The implication is that the European economy needs to be more high-tech, that is, policy should support the development and growth of high technology sectors. Given that specialization in most European economies is in low technology sectors, and given the differences in economic development across EU member states, it is questionable whether an undifferentiated high-tech strategy is the right formula to make Europe more competitive. This chapter provides some indications about how to address this question.

We should first consider the current state of economic development of individual EU member states. The evidence in the first part of the chapter shows that there are considerable differences in national technological profiles. More economically well developed member states generally have higher indigenous R&D investment than catching-up countries such as the NMS. However, even this distinction is too rough. Our analysis shows that NMS should be further differentiated. Hungary, Estonia, the Czech Republic and, to a lesser extent, Slovakia are in a process of economic catch-up that is related to the acquisition of foreign technology. This does not apply so much to the other NMS, such as Bulgaria, Romania, Latvia and Lithuania, where the catching-up process is still very much led by factors other than technological innovation.

High levels of business R&D investment require high levels of up-front public investment in the innovation system, most importantly in the education and research infrastructure. However, the effects of these investments in part are cumulative and become effective only after several years. Also, in part, it is in the nature of things that less wealthy countries cannot afford to invest large sums in this part of their economies and especially if the investment is likely to show returns only after a considerable time lag. Since the competence base in these countries is not sufficient to meet the requirements of technology-intensive innovative enterprises they have to rely heavily on technology from abroad. The case of Hungary, Estonia and especially Ireland, coupled with the post-world war II experience in many European countries, shows that the way to greater wealth is technology transfer. This, of course, implies different institutions from those required by countries with high investment in research.

Policy-makers also need to be aware that in considering policy mixes for more innovation and competition, the country's industrial structure or specialization is crucial. Industrial structure plays an important role, although in varying degrees, depending on the state of development. For countries on a catching-up path, industry structure is not a huge constraint on the absorption of and investment in new technologies and research. Most firms in catching-up countries grow and prosper on the basis of cost advantages or specific resource endowments. Their R&D levels may be suboptimal, but as the original sources of competitive advantage disappear, R&D and technological innovation become more important. Once industry-specific optimal levels of R&D are within sight, the returns from research investment start to decrease and industry structure becomes a greater constraint. At this stage, which seems to apply to most of the economies in the second group identified by the cluster analysis (AT, BE, DE, NL, UK, FR, NO, DK), countries find that structural change towards technology-intensive industries leads to changes in their technological profiles. Finally, the examples of Sweden and other countries that are technology leaders, show that if a technology intense industry structure and appropriate institutions are in place, then the constraints imposed by an industrial structure shift, and more technology can be produced and absorbed.

Our results show that country–industry interactions are significant and, therefore, that the determinants of innovation vary across industry and country groups as their technological profiles change. Knowledge creation through R&D is not the most important driver of innovation in all cases. In countries that are not specialized in technology-intense industries, firms rely more on technologies developed in other sectors and concentrate their innovation efforts on innovation activities that are not R&D-based, such as design, marketing and learning about upstream technological innovations and absorbing them. This applies to innovation-intensive firms in the group of Southern European countries.

As we move down the economic development and technology intensity ladder, we see that R&D investment and other innovation expenditures are no longer the principal factor driving innovation in the innovation-intense industries. In the countries with these types of industries, technology transfer activities are more important drivers of innovation along with non-R&D-based innovation activities. This applies to the group of NMS that are technologically more advanced (HU, EE, CZ, SK). In the group of less advanced NMS, technology transfer is the only significant source of knowledge for product innovations.

When technology transfer is important, knowledge is acquired through capital investment in equipment, through cooperation with universities, by acquiring patents and licences and via knowledge spillovers from other firms. This indicates that the nature of the innovative output will differ. For the NMS, the evidence suggests that the firms in these countries are operating in innovation-intensive sectors, but in less innovation-intensive stages of the production chain.

It would be wrong to suggest that innovation success is based solely on R&D activity. Innovation policies based on this misperception are likely not only to miss a large number of innovators, but also to fail to encourage companies to become innovators in sectors where R&D plays a subordinate role. However, this is not to reduce the importance of investment in research: there is clearly an important direct relation between innovation and economic performance. High innovation-intensive sectors are generally more productive and show higher employment growth and better innovation performance in terms of output, than sectors with fewer R&D performers. However, this chapter shows that we need to understand the factors driving innovation within a wide perspective in order to capture this complex phenomenon and to promote innovation in industries that are not considered high-tech. We show also that these factors vary as countries and their firms develop. Research investment is important for firms close to the technological frontier, but less so for firms farther away from it.

Policy needs to take account of these differences and not to measure all countries and all industries using one yardstick: one size does not fit all. However, EU policy in particular often seems to overlook this fact. Despite its embodied flexibility, the open method of coordination does not seem to provide the right incentives for transforming national innovation systems, in line with the development and specialization of member states. This is due in part to the widespread use of a few key indicators, such as the 3 per cent goal discussed in the introduction to this chapter, and the EIS. Both are useful instruments to rally support for innovation-based growth policies; however, when used as the basis for formulating policy they need to be exploited with caution. Policies for innovation-based growth should aim at creating environments in which innovative firms can prosper, but without targeting specific industries. This implies that factual innovation policy should not be limited to STI incentives, and calls for complementary measures such as social or employment policies. It calls also for a differentiated approach that discriminates among differences in economic development and specialization. The Lisbon Strategy for Growth and Jobs is sufficiently flexible to allow such a differentiated approach.

Policies targeting high-innovation intensity are likely to fail in countries that have unfavourable innovation environments and industry structures dominated by sectors that are not knowledge-intensive. In these countries, it would be advisable to foster technology transfer, increase absorptive capacity through R&D and training, and improve the basic institutional conditions that encourage growth. We believe this captures the situation for the group of economically less advanced NMS. On the other hand, in countries where industries rely on knowledge-intensive foreign imports, innovation policy should support a gradual move from knowledge acquisition to knowledge production. Obviously, the number of science and technology graduates, provision of vocational training, and good quality research at universities will become increasingly important. This applies particularly to the group of more advanced NMS. Finally, in countries that are at or close to the technology frontier, the creation of basic conditions that are conducive to more and better research within companies and also at universities, should be at the core of innovation and science policies.

Certainly, there are some important innovation challenges across countries and industries: Reinstaller and Unterlass (2008a), for instance, argue that the most urgent innovation challenges across all countries and sectors are shortage of high-skilled human resources, and lack of venture capital and finance for business R&D. The results in this chapter and others in this volume indicate that with increasing innovation intensity IPR will become more important. This calls for the development of a European patent, which would reduce the costs associated with applying for patents in Europe. Our results indicate also that support for the entry of innovative firms is important. We show that small and new firms are generally more innovative than large and old ones, but this depends also on the technological profile of each industry. However, the consistency of these results across countries and industry groups indicates that more should be done to support new innovative firms to develop and enter the market.

Policy mixes for greater innovation and competitiveness will be successful if they are able to address the complexity of the innovation process and its changes over time. As countries develop and the structure of the economy changes, the institutions supporting innovation-based growth and competitiveness need continually to be questioned, and adapted accordingly.

NOTES

1. In general, value added is preferred since gross output distorts the weights because of the intermediate outputs contained in this measure. Weighting with value added is not possible for these data since we do not have coherent series on value added for all sectors or all countries. Reinstaller and Unterlass (2008b) use OECD ANBERD data and weight them with STAN value added shares for a partly overlapping set of countries. While the resulting values are different, the results in this chapter capture the most salient features of the analysis based on OECD data. Community Innovation Survey-4 (CIS-4) sector data are used because of their coverage of NMS.
2. The results in Reinstaller and Unterlass (2008b) suggest that the result for Belgium may be due to the poor quality of the aggregated figures for sectoral R&D spending derived from CIS-4 data.
3. CIS-4 data do not allow us to calculate values for Finland and Slovenia.
4. See Knell (2008) for technical details and data issues.
5. For more information on the project and its main reports see http://www.europe-innova. eu/web/guest;jsessionid=38E2D4FDACFED190B818CA6758A3556B#.

REFERENCES

Acemoglu, D., P. Aghion and F. Zilibotti (2002), 'Distance to frontier, selection, and economic growth', *NBER Working Paper No. 9066*, Washington, DC: National Bureau of Economic Research.

Aghion, P. and P. Howitt (2006), 'Appropriate growth policy: a unifying framework', *Journal of the European Economic Association*, **4**, 269–314.

Aghion, P., T. Fally and S. Scarpetta (2007), 'Credit constraints as barriers to the entry and post-entry growth of firms', *Economic Policy*, **22**, 731–79.

Aghion, P., D. Hemous and E. Kharroubi (2009), 'Credit constraints, cyclical fiscal policy and industry growth', *NBER Working Paper 15119*, Washington, DC: National Bureau of Economic Research.

Aghion, P., N. Bloom, R. Blundell, R. Griffith and Howitt, P. (2005), 'Competition and Innovation: an inverted U-relationship', *Quarterly Journal of Economics*, **120**, 701–28.

Bartelsman, E. and B. van Ark (2004), 'Fostering excellence: challenges for productivity growth in Europe', background document for the Informal Competitiveness Council, Maastricht 1–3 July, Netherlands Ministry of Economic Affairs.

Breschi, S., F. Malerba and L. Orsenigo (2000), 'Technological regimes and Schumpeterian patterns of innovation', *Economic Journal*, **110**, 388–410.

European Commission (2002), 'More research for Europe. Towards 3% of GDP', Brussels, 11 September, Communication from the Commission, COM(2002) 499 final.

European Commission (2005), 'Implementing the Community Lisbon Programme: a policy framework to strengthen EU manufacturing – towards a more integrated approach for industrial policy', Brussels, 5 October, Communication from the Commission, COM(2005).

European Commission (2007), 'Mid-term review of industrial policy. A contribution to the EU's Growth and Jobs Strategy', Brussels, 4 April, Communication from the Commission, COM(2007) 374.

European Commission (EC) (2008), *European Innovation Scoreboard 2007. Pro Inno Europe – Inno Metrics February 2008*, Brussels: EC, DG Enterprise and Industry.

Falk, M. (2007), 'Determinants of sectoral innovation performance', *Europe Innova Sectoral Innovation Watch Deliverable WP4*, Brussels: EC, available at http://www.europe-innova.eu/c/document_library/get_file?folderId=24913&name=DLFE-2641.pdf, accessed 2 March 2010.

Harberger, A. (1998), 'A vision of the growth process', *American Economic Review*, **88**(1), 1–32.

Hölzl, W. and A. Reinstaller (2007), 'The impact of technology and demand shocks on structural dynamics. Evidence from Austrian manufacturing', *Structural Change and Economic Dynamics*, **18**(2), 145–66.

Klevorick, A.K., R.C. Levine, R.R. Nelson and S. Winter (1995), 'On the sources and significance of interindustry differences in technological opportunity', *Research Policy*, **24**(1), 185–205.

Knell, M. (2008), 'Embodied technology diffusion and intersectoral linkages in Europe', *Europe Innova Sectoral Innovation Watch deliverable WP4*, Brussels: EC, available at http://www.europe-innova.eu/c/document_library/get_file?folderId=24913&name=DLFE-2646.pdf, accessed 2 March 2010.

Lazonick, W. (2002), 'Innovative enterprise and historical transformation', *Enterprise and Society*, **3**, 3–47.

Levine, R.C., A.K. Klevorick, R.R. Nelson and S. Winter (1987), 'Appropriating the returns from industrial research and development', *Brooking Papers in Economic Activity* **3**, 783–832.

Malerba, F. (2002), 'Sectoral systems of innovation and production', *Research Policy*, **31**(2), 247–64.

Malerba, F. (2004), *Sectoral Systems of Innovation. Concepts, Issues and Analysis of Six Major Sectors in Europe*, Cambridge: Cambridge University Press.

Malerba, F. and L. Orsenigo (1995), 'Schumpeterian patterns of innovation', *Cambridge Journal of Economics*, **19**(1), 47–65.

Nelson, R.R. (1994), 'The coevolution of technology, industrial structure and supporting institutions', *Industrial and Corporate Change*, **3**(1), 47–64.

Nelson, R.R. and B.N. Sampat (2001), 'Making sense of institutions as a factor shaping economic progress', *Journal of Economic Behavior and Organization*, **44**(1), 31–54.

Nelson, R.R. and E.N. Wolff (1997), 'Factors behind cross-industry differences in technical progress', *Structural Change and Economic Dynamics*, **8**, 205–20.

OECD (2007), *OECD Science, Technology and Industry Scoreboard 2007*, Paris: OECD.

Pakes, A. and M. Schankerman (1984), 'An exploration into the determinants of research intensity', in Z. Griliches (ed.), *R&D Patents and Productivity*, Chicago: University of Chicago Press, pp. 209–33.

Papke, L.E. and J.M. Wooldridge (1996), 'Econometric models for fractional response variables with an application to 401(k) participation rates', *Journal of Applied Econometrics*, **11**(4), 619–32.

Pavitt, K. (1984), 'Sectoral patterns of technical change: towards a taxonomy and a theory', *Research Policy*, **13**(6), 343–73.

Peneder, M. (2007), 'Entrepreneurship and technological innovation. An integrated taxonomy of firms and sectors', *Europe Innova Sectoral Innovation Watch Deliverable WP4*, Brussels: EC, available at http://www.europe-innova.eu/c/document_library/get_file?folderId=24913&name=DLFE-2642.pdf, accessed 11 March 2010.

Reinstaller, A. and F. Unterlass (2008a), *What is the Right Strategy for More Innovation in Europe? Drivers and Challenges for Innovation Performance at the Sector Level*, Sectoral Innovation Watch Synthesis Report, Luxembourg: Office for Official Publications of the European Communities, available at http://www.europe-innova.eu/web/guest/home/-/journal_content/56/10136/24914, accessed 2 March 2010.

Reinstaller, A. and F. Unterlass (2008b), 'Forschungs und Entwicklungsintensität im österreichischen Unternehmenssektor. Entwicklung und Struktur zwischen 1998 und 2004 im Vergleich mit anderen OECD-Ländern', *WIFO Monatsberichte*, 2/2008, pp. 133–47.

Robson, M.J., M. Townsend and K. Pavitt (1988), 'Sectoral pattern of production and use of innovations in the UK: 1945–1983', *Research Policy*, **17**(1), 1–14.

Sandven, T. and K. Smith (1998), 'Understanding R&D intensity indicators – Effects of differences in industrial structure and country size', *IDEA Paper No. 14*, Oslo: STEP Group.

Sapir, A., P. Aghion, G. Bertola, M. Hellwig, J. Pisani-Ferry, D. Rosati, J. Vinals and H. Wallace (2003), 'An agenda for a growing Europe: making the EU system deliver', Report of an Independent High Level Group established on the initiative of the President of the European Commission, July, Brussels.

Smith, K. (1999), 'Industrial structure, technology intensity and growth: issues for policy', paper presented at the DRUID Conference on National Innovation Systems, Industrial Dynamics and Innovation Policy, Denmark, 9–12 June.

van Pottelsberghe, B. (2008), 'Europe's R&D: missing the wrong targets?' *Bruegel Policy Brief Issue 2008/03*, Brussels: Bruegel.

APPENDIX

Table A4.1   The sector classification based on innovation intensity

NACE	Industry	Original grouping	Grouping used in this chapter
29	Machinery, nec	High	Sectors with medium to high innovation intensity
30	Computers, office machinery	High	
31	Electrical equipment, nec	High	
32	Communication technology	High	
33	Precision instruments	High	
72	Computer services	High	
73	Research & development	High	
17	Textiles	Med-high	
23	Ref. petro., nucl. fuel	Med-high	
24	Chemicals	Med-high	
25	Rubber and plastics	Med-high	
26	Mineral products	Med-high	
27	Basic metals	Med-high	
34	Motor vehicles, -parts	Med-high	
35	Other transport equip.	Med-high	
64	Post, telecommunications	Med-high	
20	Wood, -products, cork	Medium	
21	Pulp/paper, -products	Medium	
28	Fabricated metal products	Medium	
36	Manufacturing nec	Medium	
62	Air transport	Medium	
65	Financial intermediation	Medium	
74	Other business services	Medium	
10	Mining: coal, peat	Med-low	Sectors with low innovation intensity
11	Mining: petroleum, gas	Med-low	
15	Food prod., beverages	Med-low	
16	Tobacco products	Med-low	
22	Publishing, reproduction	Med-low	
40	Electricity and gas	Med-low	
41	Water supply	Med-low	
66	Insurance, pension funding	Med-low	
14	Mining: other	Low	
18	Wearing apparel, fur	Low	
19	Leather,-products, footwear	Low	
37	Recycling	Low	
51	Wholesale trade	Low	
60	Land transport, pipelines	Low	
61	Water transport	Low	
63	Aux. transport services	Low	
67	Aux. financial services	Low	

Source: CIS-3 data. For details see Peneder (2007).

Table A4.2   Variable definition for regression analysis

Dependent variable: share of turnover from products new to market, TurnMar (0...1)

Appropriability

Strategic protection	Dummy = 1 if ProSec, ProDes,ProTiM = 1
Formal IPR protection	Dummy=1 if ProReg, ProCP,ProTM = 1
Patents	Dummy=1 if OPT=1
All methods	If all of above = 1

Opportunity

Total Innovation exp. share > 5% turnover	If Rtot/Turn*100 > 5
Technology transfer	Dummy = 1 if Rmac, Roek = 1
Non-R&D related	Dummy = 1 if RtR, Rmar, Rpre =1

Cumulativeness of knowledge

Cumulativeness	See Box 4.1, industry classification

Knowledge sourcing

Internal sources, high importance	Dummy = 1 if SENT = High
Equipment suppliers, high importance	Dummy = 1 if SSUP = High
Clients and customers, high importance	Dummy = 1 if SCLI = High
Competitors, high importance	Dummy = 1 if SCOM = High
Universities, high importance	Dummy = 1 if SUNI = High
Government labs, high importance	Dummy = 1 if SGMT = High
Professional conferences, high importance	Dummy = 1 if SPRO = High
Exibitions, fairs, high importance	Dummy = 1 if SEXB= High

Entrepreneurial strategy

New corporate strategy	Dummy = 1 if ACTSTR = 1
New management techniques	Dummy = 1 if ACTMAN= 1
Change organization	Dummy = 1 if ACTORG= 1
New marketing strategy	Dummy = 1 if ACTMAR= 1
Change in design of products	Dummy = 1 if ACTAES= 1

Firm controls

Cooperations	Dummy = 1 if CO = 1
Newly established	Dummy = 1 if EST = 1
Merger >10%	Dummy = 1 if TurnInc = 1/Dummy = 1 if TurnDec= 1
Internat. market	Dummy = 1 if SIGMAR = 3
National market	Dummy = 1 if SIGMAR = 4
Firm size quantile	Quartiles 25-50%, 50-75%, 75-100% turnover of all firms in country *j*

Note: Variables refer to the CIS-3 questionnaire.

5. EU innovation policy: one size doesn't fit all!

Alasdair Reid

5.1 INTRODUCTION

The rationale for public policy intervention to support science, technology and innovation (STI) has been traditionally based on the concept of market failures.[1] Such failures occur when market mechanisms (assumed in the 'neoclassical' economics to equate with perfect competition and rational expectations) are unable to secure long-term investments in innovation due to uncertainty, indivisibility and non-appropriability of the innovation process (Arrow, 1962). Typically, a market failure manifests itself in an insufficient allocation of funding by enterprises for risky and innovative investments and, hence, the market failure approach leads to instruments that allocate resources to firms (R&D grants or tax incentives).

Since the early 1990s, STI policy-makers have increasingly begun to adopt the language of national systems of innovation (NSI) (OECD, 1997) theory, which stresses that the flows of technology and information among people, enterprises and institutions are at the core of the innovation process. Innovation and technology development are the result of a complex set of relationships among the actors in the system, which includes enterprises, universities and government research institutes. For policy-makers, an understanding of the NSI can help to identify leverage points for enhancing innovative performance and overall competitiveness.

Moreover, innovation systems theory implies that policy measures that only aim to correct market failures will not result in optimal innovation performance. Rather the primary role of the state in terms of innovation policy is to facilitate the emergence of well-functioning innovation systems (Metcalfe, 2005). As systems are defined by components interacting within boundaries, policy should seek to address missing components, missing connections and misplaced boundaries, or what can be termed system failures. Hence, the public sector role is not to promote individual innovation events, rather it is about 'setting the framework conditions' in which innovation systems can better self-organize across the range of activities in an economy and, thereby, enhance innovation opportunities and capabilities. Similarly, Rodrik (2004) argues that instead of viewing industrial policy as an outcome ('picking winners') supported through a range of instruments (subsidies, tax incentives, and so on), it is a process whereby the state and private sector jointly arrive at a diagnosis about the sources of blockages in new economic activities and propose solutions to them.

Accordingly, innovation policy needs to take account of various forms of system failure (Smith, 2000). Arnold (2004) differentiates four types of system failure:

- capability failures: inadequacies in the ability of companies to act in their own best interests, for example, through managerial deficits, lack of technological understanding, learning ability or 'absorptive capacity';
- institutional failures:[2] inadequacies in other relevant NSI actors such as universities, research institutes, patent offices and so on. Rigid disciplinary orientation in universities and consequent inability to adapt to changes in the environment is an example of such a failure;
- network failures: problems in the interaction among actors in the innovation system, such as inadequate volume and quality of links, 'transition failures' and 'lock-in' failures (Smith, 2000), as well as problems in industry structure such as too intense competition or monopoly;
- framework failures: shortcomings of regulatory frameworks, intellectual property rights (IPR), health and safety rules, and so on. These failures also concern social values such as the consumer demand, culture and social values (Smith, 2000).

Tsipouri et al. (2009) argue that deficiencies in the system of governance (policy-making, evaluation, learning processes) are a fifth form of system failure, which they term 'policy failures'. Hence, the difference in the capacities and effectiveness of governance in a country can be expected to influence, positively or negatively, overall innovation system performance.

A second element influencing the design of innovation policy is the stage of development of a national (or regional or sectoral) system. Rostow (1960) argues that economies move through 'stages of development' and that the factors driving each stage (and hence the public policies brought to bear on the actors in the economy) differ. The view of economic development as a step-by-step progression or of a simplified dichotomy between low(er) and high(er) income countries, has been challenged by a range of authors (notably in the field of development economics). Lin (2010) notes that economic development is now viewed as a continuous process or on a spectrum from a low, traditional, subsistence agrarian stage, through various middle-income, industrial stages to a modern, high-income, post-industrial stage. In this context, a main driving force of structural change is the change in endowment structure from a relatively low capital to labour ratio to a relatively high capital to labour ratio. Hence, the optimal economic structures are different at different stages of development. This applies to a country's industrial, technological, financial, legal and other structures.

The Global Competitiveness Report (WEF, 2009), adopting Porter's (1990) work on determinants of competitiveness, classifies countries into factor-driven economies, efficiency-driven economies and innovation-driven economies. At each stage the basic requirements change, with the last stage performance depending on business sophistication and innovation. According to the WEF (2009), almost all of the EU27 member states are already in the most advanced 'stage 3 innovation-driven' status, or in other words close to the technological frontier. Only Latvia, Lithuania, Poland and Romania are classed in the transition from stage 2 'Efficiency driven' to stage 3, while Bulgaria is the only European Union (EU) member state classed as stage 2. However, even among those EU members classed as stage 3 innovation-driven, there are clear differences in performance.

The classification of EU27 countries by the annual European Innovation Scoreboard (EIS) provides a more differentiated breakdown. The 2008 EIS (MERIT, 2009) splits the EU27 member states into four groups:

- Sweden, Finland, Germany, Denmark and the UK are innovation leaders;
- Austria, Ireland, Luxembourg, Belgium, France and the Netherlands are innovation followers;
- Cyprus, Estonia, Slovenia, Czech Republic, Spain, Portugal, Greece and Italy are moderate innovators;
- Malta, Hungary, Slovakia, Poland, Lithuania, Romania, Latvia and Bulgaria are catching-up countries.

The EIS groupings could be criticized for suggesting that countries should seek to move between categories to attain a similar innovation performance to a top performer. Clearly, there is some expectation, as testified to by the term 'catching-up countries', that differentiated policies tailored to specific NSI will lead to a convergence in performance. However, there is no explicit claim made by the EIS authors that the methodology used for the benchmarking is anything other than descriptive. Indeed, one of the criticisms aimed at the EIS is that it is not based on a normative model.

Hence, rather than considering the EIS groupings as stages towards some ideal norm, the argument in this chapter is that they underline that there is no single optimal model. Accordingly, there is a need to dig deeper to understand how differences in innovation systems, micro- and macro-economic conditions and the policy mix (Nauwelaers, 2009),[3] influence innovation performance and ultimately competitiveness. The performance of specific NSI is clearly influenced by the sectoral composition of the economy. Malerba (2005) suggests that specific sectors in an economy face different challenges due to different market structures and industry dynamics, that innovation is driven by different factors (sources of innovation, demand, and so on), that technological regimes (the knowledge and learning environment in which firms operate) and innovation processes or modes are different.[4] Indeed, innovation processes differ greatly from sector to sector in terms of development, rate of technological change, linkages and access to knowledge, as well as in terms of organizational structures and institutional factors. Some sectors (notably high-technology ones) are subject to rapid change and radical innovations, others by smaller, incremental changes through diffusion of technologies or organizational change (OECD, 2005; Malerba, 2005).

Certainly, the sectoral structure of national economies appears to play a decisive role in terms of the trends and intensity of business expenditure on research and development (BERD).[5] The role of the structure of the economy (structural effect or 'between-industry' effect) appears to be a more important explanatory factor than sector-specific R&D intensities (intrinsic effects) in explaining differences between the EU27 and US rates of R&D investment (Duchêne et al., 2009). Such differences in innovation activity across sectors place different demands on the organizational structure of firms, and institutional factors such as regulations and IPR can vary greatly in terms of their role and importance. Hence, differences in the sectoral composition of economies need to be considered when designing an 'optimal' innovation policy.[6]

Moreover, Aghion (2006) argues that Europe's lagging performance in growth and productivity is partly a reflection of the need to invest more in innovation as the gains from capital accumulation and technological imitation were exhausted in the last quarter of the twentieth century. He notes, in particular, that the survival and growth of all sectors (from textiles to pharmaceuticals) in a high-cost, high-productivity economy depend on their ability to innovate. Following this line of reasoning, the Barcelona target (or at least the aim to increase R&D expenditure) is relevant since as EU countries move closer to the world technological frontier, they should invest more in R&D and, within the EU, the most advanced countries should invest proportionally more as they benefit from more productive R&D.[7]

Indeed, Aghion (2006) argues that government intervention to support R&D and innovation will be ineffective if the basic micro- and macro-economic conditions for innovation-based growth are not in place, namely: i) competition and market entry; ii) investment in higher education; iii) reform of credit and labour markets; and iv) countercyclical fiscal policy.[8]

Similarly, Metcalfe (2005) argues that the systems failures approach takes for granted the significance of an economic climate, with low real interest rates and stable macro-economic, micro-economic (for example the role of competition policy in ensuring that resources are distributed to more innovative firms) and monetary conditions that encourage investment in all forms.

Both the systems failure concept and arguments about the importance of taking into account the closeness to the technological frontier imply that it is not enough to invest more in R&D to get the economy to grow faster. Rather, there is a need to place an emphasis on what could be termed, in line with Metcalfe's reasoning, a 'broad-based' innovation policy. A broad-based innovation policy does not refute the need to invest in 'excellent science' or tackle system failures leading to suboptimal investment in formal R&D. However, it acknowledges that this is not sufficient, even in the NSI closest to the technological frontier, to improve innovation capabilities and performance.

A broad-based innovation policy can be considered from several standpoints. First, it is about shifting the focus from 'knowledge generation' to knowledge diffusion through policy measures targeting the diffusion of technologies, but also organizational methods in firms (and indeed in society). In this approach, innovation policy interacts strongly with education and training policies as well as integrating support for improved engineering, software, business management and of course technology take-up (notably ICT).

Secondly, a series of theoretical and empirical approaches equally places greater emphasis on the 'demand side' of the innovation process, underlining that the sources of inspiration for innovations are, as often, drawn from consumers, co-developed by users (von Hippel, 1988), influenced by standards-setting processes to target lead markets, and boosted by public procurement (Edler, 2005; Georghiou, 2007). These strands of thinking have rapidly found their way into policy, notably through the impulsion given by the Aho (2006) report. This report argues that the reason business is failing to invest enough in R&D and innovation in Europe is the lack of an innovation-friendly market in which to launch new products and services. To create such a market requires action for: better regulation; ambitious use of standards; driving demand through public procurement; a competitive IPR regime; a culture that celebrates innovation.

Much of the recent literature on innovation policy underlines the need to shift to a horizontal (or cross-cutting) form of policy intervention, the so-called 'third-generation' innovation policy (Legrand et al., 2002). Innovation policy, hence, transcends traditional sectoral (or vertical) policies such as competition, education and training, environmental, transport, health, and so on. The MONIT (monitoring and implementing national innovation policies) study (OECD, 2005) takes this thinking further by distinguishing between narrow innovation policy, aimed primarily at innovative firms (essentially an extension of classic industrial and technology policies), and innovation policy in a wider sense, aimed at economic development and quality of life, requiring an 'innovation everywhere' approach and a coherent policy-mix with linkages between 'narrow' innovation policy and innovation in other policy domains.

In summary, this chapter adopts the hypothesis that, depending on the sectoral composition of an economy and its closeness to the technological frontier, the overall level of economic development and the type and extent of failures in innovation systems, different EU member states need to adopt differentiated 'innovation policy mixes' in order to boost their innovation performance. In all cases, the policy mix needs to be broad-based, irrespective of the 'stage' of development, since failing to take into account the 'framework conditions', such as education, property rights and finance, and competition, will undermine the likely effectiveness of more targeted measures.

To explore this assumption, this chapter examines the extent to which innovation policy in five Central and Eastern Europe countries (CEEC5) is relevant to the specific strengths and weaknesses of their NSI. The five countries, Czech Republic, Hungary, Poland, Slovakia and Slovenia, have been selected since they represent a specific case of 'transition' or 'catching-up' countries in the wider EU context. This chapter examines the challenges faced in designing innovation policies tailored to these EU members. The following sections:

- consider briefly recent trends in the specific innovation performance and capabilities of the CEEC5;
- discuss the importance of the types of 'system failure' in the NSI of the CEEC5 and to what extent they are addressed by national policy;
- review the innovation policy mix, including the role of EU Structural Funds, in the CEEC5 and the correspondence with the system failures identified.

In examining the innovation performance and policy of the CEEC5, this chapter draws on work undertaken, during the decade since 1999, in the framework of the European TrendChart on Innovation and associated projects (notably the EIS). The author duly acknowledges the work done by many colleagues on the TrendChart project, including those from the CEEC5, while taking full responsibility for any conclusions drawn here.

5.2   MAIN CHALLENGES FOR INNOVATION POLICY: EU27 VS CEEC5

5.2.1   Catching up? Recent Innovation Performance of CEEC5 Countries

What implications for innovation policy can we draw from recent work at national or European level with respect to the specific capabilities and performance of the NSI in the CEEC5? Six key elements are worth highlighting.

First, Figure 5.1 presents the re-scaled scores for the CEEC5 vis-à-vis the EU27 for the five main groups of EIS 2007 indicators, where the minimum value is 0 and the maximum value 1.[9] The CEEC5 perform below the EU27 except for the Czech Republic and Slovakia for applications, and Slovenia for innovation drivers. Broadly speaking, Slovenia performs best amongst the CEEC5 for knowledge creation while Slovakia

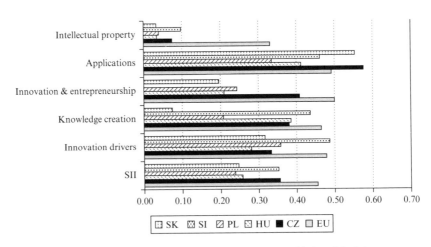

Note: No value for Slovenia for innovation and entrepreneurship in original dataset.

Source: Adapted from MERIT (2007).

Figure 5.1   Comparative innovation performance CEEC5 vs EU27 (2007)

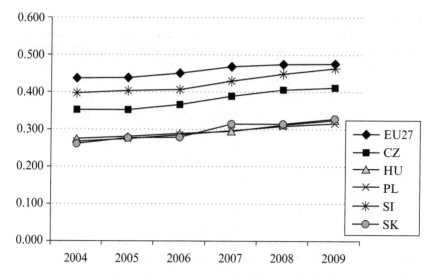

Source: Author based on MERIT (2010).

Figure 5.2   Trends in the summary innovation index 2004–09, CEEC5 vs EU27

is the worst performer. The Czech Republic performs best for innovation and entrepreneurship.

The most obvious common characteristic of the CEEC5 is the very weak performance in intellectual property. A number of observers have pointed out that intellectual property indicators are not the most relevant for tracking these economies. For instance, Havas (2008) argues that for a 'catching-up' economy it is not meaningful (or feasible) to produce a high number of patentable R&D results. It is more relevant to concentrate on fostering the diffusion of new technologies, and enhancing capabilities for more efficient absorption of new methods and technologies. However, weak performance in intellectual property does underline that the NSI of the CEEC5 remain relatively far from the European and, hence, global, technology frontier. In a recent analysis of regional scientific and technological specialization patterns, Peter and Frietsch (2009) found that only one region in the CEEC5 made it into the top 100 patenting regions of the EU27 (the Budapest region).

A second point is that while all the CEEC5 have been making progress (based on their summary innovation index), they all remain below and in some cases far below even average EU performance levels (either as a whole or for certain indicators) (Figure 5.2). The time required to catch up varies, but even the best performers amongst the CEEC5 (the Czech Republic and Slovenia) are still a decade away from catching up to the EU27 average, and far from becoming 'world leaders' (EC, 2007). The latest results from EIS 2009 confirm this finding with only Slovenia showing convergence, with a summary innovation index (SII) score of 0.466 compared to the EU27 average of 0.478, while the Czech Republic is the only one of the CEEC5 classified as a growth leader in terms of the SII trend in the period 2004–2009 with an average annual growth rate of 4.8 per cent compared to 1.8 per cent for the EU27. In the three other CEEC countries, the rate of convergence is slight, with average annual growth rates barely above the EU27 rate.

A third finding should provoke some reflection on the fact that innovation policies are focused on the type of innovators active in the innovation systems: while the position of individual CEEC5s varies, the majority of enterprises in Hungary (57 per cent) and Poland (82 per cent) are 'non-R&D innovators'[10] (Arundel et al., 2008); while in the Czech Republic and Slovakia, respectively 45 per cent and 40 per cent of enterprises are also non-R&D innovators. This is in part a reflection of the industrial structure of these countries (non-R&D innovators are concentrated in low-technology manufacturing and service sectors) and the closeness of the NSI (and specific sectoral systems) to the technological frontier. The CEEC5 can be classified as technology users (see Reinstaller and Unterlass, 2008); indeed, Knell (2008) finds that the R&D content in the economies of Poland, Czech Republic and Slovakia is below the EU average. Moreover, the CEEC5 have relatively low company-specific R&D intensity and tend to rely on R&D embodied in imported inputs. Since non-R&D innovators innovate primarily through technology transfer and training, this suggests that policy in countries with high rates of non-R&D innovators should be targeted towards relatively more support for these aspects through measures that combine advice on 'best-available technologies', financial incentives to upgrade, and training programmes or centres targeting specific technological and organizational skills in the workforce.

Fourth, a common characteristic of the CEEC5 'innovation systems' derives from industrial structures and ownership dynamics, distinguishing them from the Southern European 'cohesion' countries. A notable distinction is the high penetration of foreign direct investment (FDI) in these economies (a characteristic shared with the Baltic States). Radosevic (2006) sums up the structural weaknesses of the CEEC5 innovation systems as being based on innovation activity restricted to a few large firms (with foreign-owned firms investing more than domestic companies).

A fifth characteristic of the CEEC5 economies is that while these countries generally have restructured, more or less successfully, towards medium to high-tech sectors, and while trends in BERD are generally positive, the gap between R&D and innovation intensity remains wider than the gap in employment patterns. This suggests that the majority of activity in the medium-to-high-tech sectors remains focused on production of goods rather than 'production of new knowledge', and that much of this shift is due to foreign-owned firms investing in the 'low-end' of high-tech. As Havas (2008) notes, Hungary's performance is better (close to the EU average) on only a few indicators, such as high-tech exports, share of high-tech R&D, and employment in high-tech services. However, this result is generated by the weight of foreign-owned firms in the Hungarian economy, statistically belonging to high-tech sectors, but mainly engaged in low knowledge-intensive activities.

Equally, the relative weakness of the knowledge-intensive services (KIS) sector in the CEEC5 is a cause for concern. Data (MERIT, 2009) on KIS employment and exports show that the CEEC5 economies are lagging significantly behind the EU27 average; this contrasts with the relatively good, even if nuanced by the preceding comments, position in high-tech industry employment and exports.

The sixth and final element of CEEC innovation systems is to be found in less traditional innovation indicators that offer greater insight into the range of factors influencing the social capability of innovation systems to foster change. Freeman (2006) argues that the obvious divergence, particularly in Eastern Europe, in growth rates over the long term is attributable in large measure to the presence or absence of social capability for institutional change, and especially for those types of institutional change that facilitate and stimulate a high rate of technical change. Social surveys, such as the EU's InnoBarometer, provide only a snapshot of the evidence on whether a NSI is more or less open to institutional change, technological adoption and organizational innovation. Bruno et al. (2008) bring together evidence from a range of surveys and standard statistical sources to investigate the broader 'socio-cultural factors' influencing innovation in the EU25 (minus Bulgaria and Romania).

Four 'capitals' (or dimensions) are used to identify the socio-cultural characteristics of NSI relevant to innovation: cultural capital and consumer behaviour (or demand for innovation); human capital; social capital; and organizational capital and entrepreneurship. Bruno et al. (2008) find these factors to be positively correlated to standard innovation or competitiveness indicators such as intensity of BERD and labour productivity.

The overall findings are summed up in Figure 5.3 with only Slovenia out of the CEEC5 performing well, and, at the other extreme, Poland

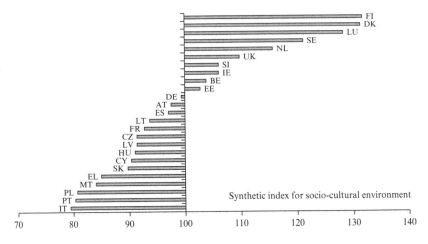

Source: Bruno et al. (2008).

Figure 5.3   Socio-cultural environment for innovation

(close to last with Portugal and Italy). The Czech Republic, Hungary and Slovakia, while doing better than Greece and only just trailing France, have rigid socio-cultural environments (under-performing for cultural, human and organizational capital). In terms of specific 'capitals', Hungary performs well for cultural/consumer capital, indicating a more open attitude towards innovation and new technologies; Slovenia scores best for social capital (indicative of a high level of trust in society, lower levels of corruption and open cooperation networks) while Poland fares particularly poorly for this category. The CEEC5 are in the worst performing group for human capital with 'rigid labour markets': the availability of higher educated people is low, as is the share of human resources in science and technology; the workforce is particularly immobile and less involved in lifelong learning, and the share of non-national human resources in science and technology is also particularly low (Bruno et al., 2008). Finally, in terms of organizational capital, Hungary and Slovenia are among the weakest countries, with 'rigid organizations'; and the Czech Republic and Slovakia show particularly low levels of importance given to initiative at work by individuals.

5.2.2   Innovation Challenges and Priorities in the CEEC5

Given the specific characteristics of innovation performance, this section assesses the match between the innovation policy challenges, on the one hand, and the innovation policy priorities and targets set by the national governments in the CEEC5, on the other.

Column 2 in Table 5.1 shows for each of the countries the innovation policy challenges identified in the TrendChart countries for the period 2006–2008.[11] The challenges for the CEEC5 can be grouped as follows:

- quantity and quality (skills composition) of human resources available within the innovation system;
- improved exploitation of R&D results through increased cooperation between research base and enterprises;
- increased innovation activity within enterprises (reports generally refer to small and medium sized enterprises (SME) and/or domestic enterprises).

Clearly these common challenges are over-simplified, although the 'challenges' presented in Table 5.1 are expanded on and justified in each country report. The picture that emerges is in part coherent with the preceding analysis, notably in terms of the relative weakness of the human resources available for innovation. However, most of the reports note that this is not an issue that is easily solved since it is dependent not only on profound changes in the education system (there is a reported divergence between increasing numbers of higher education students generally, and the capacity of the education systems to produce quality graduates), but also the value systems (the social attractiveness of pursuing a scientific or technological education as opposed to social sciences or business-related studies). Equally, the above-mentioned weaknesses in socio-cultural capital imply that it is not enough simply to increase the numbers of science and technology graduates, but that education and training systems need to focus on organizational innovations and broader actions will be required to promote 'innovation awareness'.

The second common challenge assumes that scientific specialization (and quality) of each of the national public/academic research bases provides knowledge that can be used to create competitive advantage in the market. The TrendChart country reports often link weakness in patenting to lack of commercialization of research results, which assumes that the research results of universities are either relevant to the existing industrial structure of the economies, or that the universities are organized and have in place management and incentive systems allowing them to patent and then support the results through proof of concept stage to the creation of spin-offs.

The third common challenge is that 'the vast majority of R&D/innovation is done by large/foreign-owned enterprises while most smaller

Table 5.1   CEEC5 innovation challenges and priorities

Country	Innovation policy challenges (2006–2008)	Innovation policy priorities/targets
Czech Republic	Weak cooperation between universities/ public R&D and industry as evidenced by low patenting activity; Financing of research, development and innovation, notably innovative (risky) start-ups; Human resources for the knowledge economy.	National Innovation Policy (2005–10) has four broad strategic objectives: − strengthen R&D as a source of innovation; − establish working cooperation between the public and private sector; − secure sufficient human resources for innovation; − make public administration in research, development and innovation more effective.
Hungary	Low share of innovative firms; Low occurrence of cooperation in innovation activities; Potential gaps in the quantity and quality of human resources for RTDI.	STI Strategy (2007): − culture of embracing and exploiting S&T results; − quality-, performance- and exploitation-driven, efficient NIS; − respected, creative and innovative workforce suited for the needs of the 'knowledge-based' economy and society; − legal and economic environment stimulating the creation and utilization of knowledge; − indigenous businesses that are considered competitive on the global markets.
Poland	Stimulate and deepen innovation internal capacities of Polish companies; Improve science– industry cooperation; Promote multidisciplinary profile of skills.	Innovation Strategy (2006) – five priorities: − human resources for modern economy; − applied research; − intellectual property; − access to finance; − infrastructure for innovation.

Table 5.1 (continued)

Country	Innovation policy challenges (2006–2008)	Innovation policy priorities/targets
Slovakia	Low shares of innovative enterprises limit competitiveness of the country; Low spending on R&D generates extremely low commercial output of the Slovak research base; Lacking human resources for R&D system and weak ties between industry and academia sectors.	Innovation Strategy (2007): – High-quality infrastructure and an efficient system for the development of innovations; – High-quality human resources; – Efficient innovation policy tools.
Slovenia	Better exploitation of R&D results and closer links between public R&D and business sector; Increase innovation activity, especially in SMEs; Development of human resources to support innovation activity.	National Research and Development Programme (NRDP) (2005): – Main priority is the need to increase even more dynamically the level of investment of the private sector for R&D.

Source: Adapted from TrendChart country reports, 2007, 2008.

domestic firms are unable to innovate due to access to finance'. In comparing the arguments put forward to justify these challenges, the CEEC5 TrendChart reports tend to blame low levels of innovativeness on the inefficiency of public support for enterprises.

In terms of how policy-makers set objectives and targets for innovation policy, national priorities (see column 3 in Table 5.1) are taken from each country's official government 'policy documents', with the exception of Slovenia, which has no specific stand-alone innovation strategy document and the national R&D plan is used as a proxy. As highlighted in the reports, the various innovation policies or strategies are not necessarily formulated as coherent or logical strategic targets followed by operational action plans within a policy-making framework. However, it is possible to identify a common set of policy priorities across the CEEC5, namely:

- investment in human resources for innovation (except Slovenia);
- improving the quality and functioning of the 'innovation infrastructure', more specifically innovation (financial) support systems and the ICT infrastructures;
- strengthening investment in applied research and innovation in enterprises.

Somewhat counter-intuitively, only Hungary's STI strategy gives priority to the 'legal and economic environment stimulating the creation and utilization of knowledge' (Havas, 2008), although it might be expected that the legal and regulatory barriers to innovation in the CEEC5 would be greater than in, say, Western European countries.

It is noticeable that certain elements that could be considered key failures in the NSI of the CEEC5 are not taken into account in the challenges identified or in the policy priorities. These include most institutional or framework type failures that, at best, are considered 'sub-plots' of the identified challenges. Leaving aside the weaknesses in education and financial systems, there is very little in the reports about the sort of 'demand-side' or institutional failings that may hinder innovation in the NSI of the CEEC5.

Equally, the role of FDI in driving innovation and technological and organizational change is not fully addressed, despite some recognition of the importance of multinationals in innovation investment and economic restructuring (towards higher value added). For instance, Balasz (2008) notes that, in the case of Slovakia, the dual economy and weak ties between domestic SME and branches of multinationals seem to be the greatest challenge for national innovation performance. However, there is no simple indicator to measure the operations of multinationals and their role in technology transfer and the introduction of new modes of business.

Cooperation in the innovation systems is reduced to weak links between the public/academic research base (in most CEEC5 countries far from efficient or effective) and enterprises (the reports question the capabilities of enterprises to express demand for R&D services).

5.2.3 System Failures: A Comparative Analysis of the EU27 and CEEC5

This section draws on work carried out for the European Commission (EC) within the framework of the European TrendChart on Innovation Project. Tsipouri et al. (2009) adopted the 'systems failure' approach, explained previously, to analyse the rationale for public sector support for innovation. We apply this concept here to provide an analytical framework for analysing innovation policy in the EU27 in general, and the CEEC5 in particular. The six types of failures may overlap or be difficult to interpret. Indeed, policy challenges and measures often address more than one failure (see Table 5.2). For example, a measure aimed at increasing the innovation capacity of companies through technology transfer between companies and research institutes can be classified in terms of both capability and network failures.

Based on the challenges identified in the 2008 EU27 TrendChart country reports, Figure 5.4 displays the relative weightings given to the innovation policy challenges related to six types of failure. As can be seen, capability failure is the most significant, ahead of institutional and market failures. This suggests that there are significant capability failures (limited management skills, weak know-how on technological or organizational innovation, and so on) in companies which are a considerable impediment to intensifying innovation.

For many years lack of financing has been considered one of the major barriers to innovation in SME. Limited access to finance for innovation projects is classically considered a market failure. However, there are reasons to suggest that in many EU countries these failures are more institutional in nature, that is, failures in the structure or operating procedures of financial markets, or framework failures, that is, inappropriate regulatory environment to provide encouragement to financial actors to invest in riskier ventures. However, the focus of the challenges related to capability failures suggest that policy needs to focus more on supporting enterprises to improve their capabilities to manage innovation processes in-house.

Figure 5.5 distinguishes the importance of the challenges identified by EIS 2008 country groups. The CEEC5 are located in the moderate innovators (Czech Republic and Slovenia) and the catching-up countries (Hungary, Poland and Slovakia). Recall that the group of innovation leaders includes five countries, the innovation followers six, the moderate innovators eight and the catching-up countries eight. Since one challenge can address more than one type of failure, the figure shows absolute numbers rather than percentage shares.

While capability failures are perceived as important for all four groups, failures in framework conditions are considered more significant for innovation leaders and followers. This could be seen as a rather paradoxical finding, but it does not imply per se that framework conditions are weaker in these countries. Rather it is indicative that framework conditions are more often identified as a policy issue, possibly because the basic

Table 5.2   Examples of challenges per type of failure in the EU27

	Innovation leaders	Innovation followers	Moderate innovators	Catching up
Market failures	UK: Boost relatively weak intensity of innovation activity in enterprises	IE: Increase the level of innovation in the private sector	CY: Increase inputs and efficiency of business innovation	BG: Increase R&D expenditure (private and public)
Capability failures	FI: Broaden the base of innovative growth-oriented enterprises	FR: Increase non-technological innovation (organizational, design) innovation in SMEs	EE: Building competences and developing innovation management skills	PL: Stimulate and deepen innovation internal capacities of Polish companies
Institutional failures	DE: Increasing supply of highly qualified labour	BE: Innovation skills mismatch	IT: Innovation financing	RO: Improve innovation and business support infrastructure (business incubators, technology transfer offices, S&T parks, etc.)
Network failures	SE: Centres of Excellence: creation of globally competitive research and innovation milieux	LU: Reinforce synergies, complementarities and collaborations between the public and private RD centres	CZ: Cooperation between public R&D and industry	SK: Development of knowledge-intensive clusters across public knowledge poles
Framework failures	SE: Innovative public procurement: revitalizing old models to transform knowledge to commercial value	FR: Foster intellectual property use by SMEs	ES: Decreasing availability of human capital and skills	MT: Sustaining enhanced investments in business R&D and encouraging innovation of SMEs

Table 5.2 (continued)

	Innovation leaders	Innovation followers	Moderate innovators	Catching up
Policy failures	FI: Transformation of firm strategies and new innovation models	n.a.	EL: Low effectiveness and limited impact of the innovation measures on economy and employment	SK: Under-developed innovation governance

Notes: BE: Belgium; BG: Bulgaria; CY: Cyprus; CZ: Czech Republic; DE: Germany; EE: Estonia; EL: Greece; ES: Spain; FI: Finland; FR: France; IE: Ireland; IT: Italy; LU: Luxembourg; MT: Malta; PL: Poland; RO: Romania; SE: Sweden; SK: Slovakia; UK: United Kingdom

Source: Adapted from Tsipouri et al. (2009).

conditions (internal capabilities of enterprises, innovation infrastructure, access to finance or innovation support services, and so on) are already better than in the lagging countries.

It is perhaps surprising that network failures (industry–science cooperation, clustering, and so on), often highlighted in policy debates as a weakness of many NSI in the EU, are highlighted less often than capability, institutional and market failures. However, network challenges are relatively more present among the moderate innovators and catching-up countries, suggesting that innovation cooperation and knowledge transfer are more problematic in these 'less-developed' innovation systems.

Experts in the catching-up countries were particularly concerned about network failures. The Hungarian report points to 'low occurrence of cooperation in innovation activities'; the Bulgarian report calls for action to 'stimulate partnerships and to increase cooperation between science institutions, enterprises and other institutions involved in the innovative process'; the Poland, Romania and Slovakia reports highlight the need to improve industry–science cooperation; as does the Czech report (Pazour, 2008) for the group of moderate innovators.

An example of a network challenge (Walendowski, 2008) is weak science–industry linkages in Poland. This is a collateral result of the current policies encouraging research teams to publish research results rather than support market exploitation. The Polish business sector considers it difficult to build working relationships with researchers

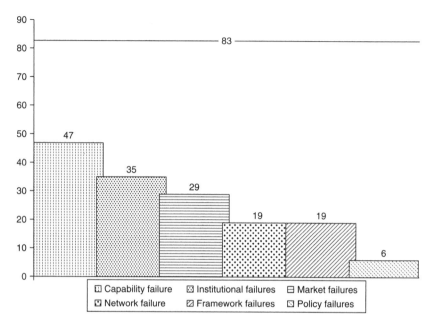

Note: The numbers over the vertical bars indicate a number of challenges addressing a given failure. One challenge can address more than one type of failure. There were 83 challenges defined in the 2008 country reports.

Source: Tsipouri et al (2009).

Figure 5.4   Failures targeted by EU27 innovation policy challenges

although, when it occurs, cooperation is considered beneficial. According to the survey of firms conducted by the Ministry of Science and Higher Education in Poland, 40 per cent of companies have never tried to establish cooperation with a research institution and more than half of the companies interviewed stated that cooperation with research institutions was not a priority. In 2006, 483 industrial enterprises had concluded cooperation agreements relating to innovation activities with branch research institutes and 540 had collaborated with higher education institutions. In nominal terms, this means that the business sector financed 15.4 per cent of branch research institute R&D investment and 5 per cent of the investment made by higher education institutions, which is an indication of the low levels of science–industry cooperation.

Policy failures are not as prominent in the analysis as might be expected, but this is due more to use of the EIS framework, which does not include specific policy or governance metrics, than a sign that all is

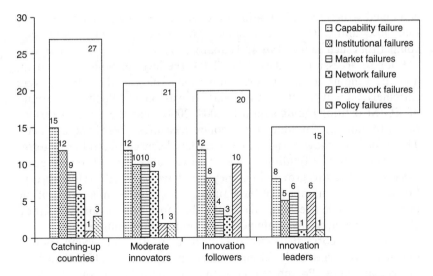

Notes: The numbers over the vertical bars refer to the number of challenges addressing a given failure. The numbers in the right upper corner of the black frames are total numbers of challenges in EIS groups. One challenge can address more than one type of failure.

Source: Tsipouri et al. (2009).

Figure 5.5 Challenges by type of failure and EIS group

well. Governance and policy management issues are dealt with in the TrendChart reports, and the weakness of the CEEC5 governance systems (one element of the social capabilities discussed above) is identified as a source of concern.

5.3 THE POLICY MIX FOR INNOVATION: EU27 VS CEEC5

This section provides a broad overview of the current policy mix in the EU27, favouring innovation and STI more broadly. It compares the patterns found in the leading EU27 countries with those for the CEEC. The analysis is based on the European Inventory of Research and Innovation Policy Measures, a platform for national information and documentation on research and innovation policies and measures. The inventory has been built up since 2000, with financial support from the EC, first through the European Innovation TrendChart project[12] and more recently in partnership with the associated ERAWATCH initiative.[13] In 2009, the inventory contained structured information on over 1200 policy measures in more than 40 countries (EU27, associated and candidate countries and main competitors such as the USA and Japan).

Prior to accession to the EU in 2004, the level of policy interest in, and the funding available for, innovation policy in the CEEC5 were extremely limited (ADE, 2001, 2003). It was only with the first round of EU Structural Funds support (2004–2006) that there was a serious boost to funding for innovation policy measures (Technopolis, 2006). The impact of these Structural Funds has been considerable in terms of increased public funding; however, governance capabilities have not always matched this growth in money, and have often led to slow absorption and uncertain impact of these funds. According to an analysis of the period 2000–2006 (Technopolis, 2006), the capacity to absorb Structural Funds for research, technological development and innovation (RTDI) appears to be correlated with overall Structural Funds spend (only 6 out of 22 countries analysed had higher absorption of RTDI than total Structural Funds; Poland is one of the six, based largely on significant spending on science park infrastructure), and by 2006 most of the new member states were still reporting very low absorption figures for RTDI-related spending programmes. Hence, despite the identified needs for support amongst firms for new technologies, product development and so on, the 'institutional bottlenecks' in the CEEC5 innovation governance systems appear to have limited the immediate effectiveness of the inflow of 'fresh cash'. Technology transfer measures (embodied innovation through technology acquisition) have experienced the slowest start despite the need to improve low rates of manufacturing productivity (in most CEEC5).

According to Tsipouri et al. (2009: 27), the policy priority most often addressed by STI measures in the EU27 member states is 'support for R&D cooperation including joint research projects run by public-private consortia of business and research' (nearly one third of all support measures, see Figure 5.6). The top ten priorities are dominated by funding for technological R&D and measures supporting research–industry linkages suggesting an ongoing bias in innovation policies towards a 'linear' or research laboratory to market approach. Measures aimed at more disruptive market-driven innovation (for example support for innovative start-ups), organizational innovation (support for innovation management and related advisory services) or building cooperation and more open innovation (cluster support) are among these priorities. Given the usual emphasis on lack of finance for innovation as the barrier to smaller firms engaging in this activity, the relatively lower priority given to the group of financial measures (risk capital, horizontal measures, fiscal incentives, guarantees)

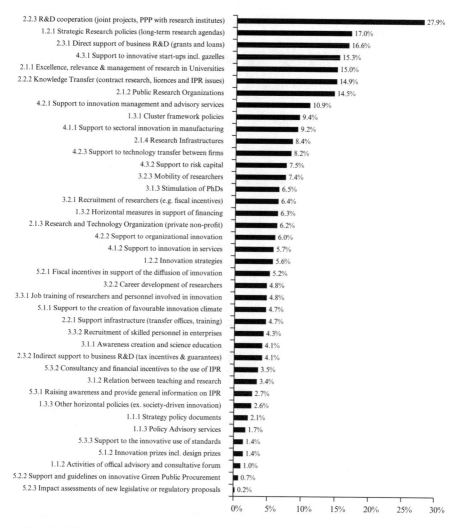

2.2.3 R&D cooperation (joint projects, PPP with research institutes) — 27.9%
1.2.1 Strategic Research policies (long-term research agendas) — 17.0%
2.3.1 Direct support of business R&D (grants and loans) — 16.6%
4.3.1 Support to innovative start-ups incl. gazelles — 15.3%
2.1.1 Excellence, relevance & management of research in Universities — 15.0%
2.2.2 Knowledge Transfer (contract research, licences and IPR issues) — 14.9%
2.1.2 Public Research Organizations — 14.5%
4.2.1 Support to innovation management and advisory services — 10.9%
1.3.1 Cluster framework policies — 9.4%
4.1.1 Support to sectoral innovation in manufacturing — 9.2%
2.1.4 Research Infrastructures — 8.4%
4.2.3 Support to technology transfer between firms — 8.2%
4.3.2 Support to risk capital — 7.5%
3.2.3 Mobility of researchers — 7.4%
3.1.3 Stimulation of PhDs — 6.5%
3.2.1 Recruitment of researchers (e.g. fiscal incentives) — 6.4%
1.3.2 Horizontal measures in support of financing — 6.3%
2.1.3 Research and Technology Organization (private non-profit) — 6.2%
4.2.2 Support to organizational innovation — 6.0%
4.1.2 Support to innovation in services — 5.7%
1.2.2 Innovation strategies — 5.6%
5.2.1 Fiscal incentives in support of the diffusion of innovation — 5.2%
3.2.2 Career development of researchers — 4.8%
3.3.1 Job training of researchers and personnel involved in innovation — 4.8%
5.1.1 Support to the creation of favourable innovation climate — 4.7%
2.2.1 Support infrastructure (transfer offices, training) — 4.7%
3.3.2 Recruitment of skilled personnel in enterprises — 4.3%
3.1.1 Awareness creation and science education — 4.1%
2.3.2 Indirect support to business R&D (tax incentives & guarantees) — 4.1%
5.3.2 Consultancy and financial incentives to the use of IPR — 3.5%
3.1.2 Relation between teaching and research — 3.4%
5.3.1 Raising awareness and provide general information on IPR — 2.7%
1.3.3 Other horizontal policies (ex. society-driven innovation) — 2.6%
1.1.1 Strategy policy documents — 2.1%
1.1.3 Policy Advisory services — 1.7%
5.3.3 Support to the innovative use of standards — 1.4%
5.1.2 Innovation prizes incl. design prizes — 1.4%
1.1.2 Activities of offical advisory and consultative forum — 1.0%
5.2.2 Support and guidelines on innovative Green Public Procurement — 0.7%
5.2.3 Impact assessments of new legislative or regulatory proposals — 0.2%

Source: Tsipouri et al. (2009).

Figure 5.6   Policy measures targeting specific priorities (EU27, 2008)

is not surprising. Measures addressing human capital seem relatively under-represented in the STI policy mix.

As might be expected, the picture changes when the policy mix is analysed at a disaggregated level with a focus on countries with different levels of development. Analysis at the level of EIS country groups reveals some substantial variance in terms of policy priorities addressed by national STI

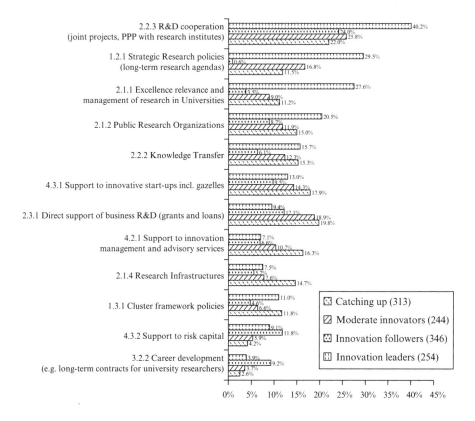

Source: Tsipouri et al. (2009). Based on TrendChart-ERAWATCH database of support
measures (N=1157).

Figure 5.7   Key policy priorities per EIS country groups

policies (see Figure 5.7).[14] While R&D cooperation is the key policy priority for all EIS groups, the focus on this priority among innovation leaders
(40 per cent of all measures) is significantly stronger than in any other
group. Similarly, the share of other priorities related closely to science
policy (for example strategic research policies, public research organizations and support for excellence and management of research in universities) is relatively much higher for the innovation leaders.

On the other hand, the group of innovation leaders focus much less on
providing direct support for business R&D (9 per cent) and support for
innovation management services (7 per cent), which are more of a priority
for the catching-up countries and the moderate innovators (respectively

for direct support to R&D 20 per cent and 19 per cent, and innovation management 16 per cent and 11 per cent). As expected, given their level of science and technology development, the catching-up countries tend to emphasize research infrastructures (15 per cent of all measures) more than any other group.

Tsipouri et al. (2009) note that innovation leaders concentrate on a smaller number of STI policy priorities than other countries, with four key priorities addressed by more than 20 per cent of their measures. Innovation followers have the most diverse policy mix in terms of priorities addressed, with one policy priority accounting for more than 15 per cent of all their measures. Moderate innovators and catching-up countries take a more horizontal approach with the focus spread evenly across different priorities. This may reflect a broader range of 'system failures', but also may be a sign of weaker governance capacities enabling more tailored interventions.

For policy effort (measured by number and budgets) in terms of specific measures responding to the various types of failures, evidence from TrendChart (see Figure 5.8) suggests that, in relative terms, network and policy failures are targeted most often by innovation leaders and followers, while capability failures are most frequently the focus for catching-up and moderate innovators.

Box 5.1 illustrates the types of measures targeting different types of failure, and gives a number of examples from the CEEC5 countries. Obviously, a policy measure may tackle more than one type of failure and so a measure addressing market failures may also be tackling network failures, and so on.

The CEEC5 countries have a relatively large number of measures for tackling market failures, including technology acquisition and industrial R&D and innovation projects. Since 2007, actions promoting clustering, national technology platforms and so on, have been developed to stimulate networking, alongside measures designed to build regional institutional depth in the two larger CEEC5 (Poland, Hungary). Fewer measures are devoted to tackling framework failures.

In order to deepen the analysis of policy priorities and build a more reliable image of public sector interventions to promote STI, the TrendChart-ERAWATCH policy measures database was analysed using budgetary data for mid-2009. The advantage of these data is that they allow analysis of the government spending on different policy priorities, which official statistics on public expenditure on R&D and on government budgetary appropriations on R&D, and so on, do not provide.[15]

The picture that is captured by the budgetary data presented in Table 5.3 differs from the prioritization frequency of policy measures presented

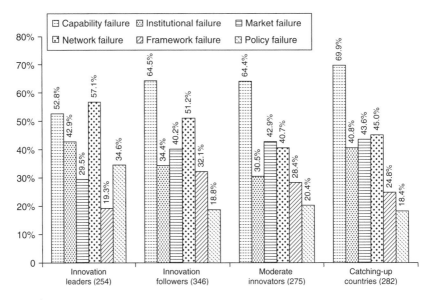

Notes: The percentages refer to the share of measures in EIS country group addressing a given failure. Measures can target more than one type of failure. The numbers in brackets indicate a total number of support measures in EIS groups.

Source: Tsipouri et al. (2009). Based on TrendChart-ERAWATCH database of support measures (N=1157).

Figure 5.8   Targeting of policy measures to system failures (EU27, 2008)

previously. Table 5.3 presents average calculated expenditure for each policy priority category as a share of the national total. The EU25 (due to incomplete data Bulgaria and Romania were excluded from the analysis) spend on RTDI policy measures in 2008 was estimated at €23.5 bn, of which CEEC5 accounted for 11.3 per cent or €2.66 bn (Poland alone accounted for just over €1 bn). Some elements common to the CEEC5 include:

- a significantly higher relative spend on technology transfer/acquisition type measures for SME in the CEEC5 than in the EU25 (18.4 per cent compared to 2.4 per cent); relatively more investment in direct support (grants and loans) for business R&D activities (15.8 per cent compared to 10.3 per cent), higher relative spending on research infrastructure (notably in Poland) (8.4 per cent to 5.6 per cent);

BOX 5.1   EXAMPLES OF MEASURES TARGETING
          DIFFERENT TYPES OF SYSTEM
          FAILURE

Measures Addressing Market Failures

Measures addressing market failures generally do so by providing financial incentives to firms to invest in R&D or innovation activities. They may seek to stimulate private sector investors to invest in riskier undertakings, e.g. through public support for early-stage investments.

Poland: Technology credit; Category 4.2.3 Support to technology transfer between firms; Period: 2007–13; Budget: €410m

The main goal is to support investment relating to the implementation of new technologies. In order to obtain the technology credit, a company must present the paid invoices of technology acquired as well as submit an independent opinion from a research organization certifying that the products/services actually were developed with the use of new technologies. The commercial banks can give unlimited loans, while the National Bank of Economy will intervene through technology credits up to PLN 4m (roughly €1.1m) upon the presentation of the paid invoices for the products/services sold as a result of the purchase and implementation of new technologies. The reason for involving the commercial banks was mainly to increase the number of loans that would be given to companies interested in the development and implementation of new technologies, while the general rationale behind this measure is to increase private R&D spending. This scheme can also be viewed as an attempt to diversify the support based primarily on grants with a new instrument that would be attractive for banks to invest in innovative undertakings.

Measures Addressing Capability Failures

Measures addressing capability failures include, for example, measures aimed at training in companies related to new technologies or organizational practices, innovation management methods, support for recruitment of skilled personnel or for young researchers to undertake PhDs, etc.. This can include indirect support via innovation centres, networks of advisors, etc.

Slovenia: Young Researchers from business sector; Category 3.2.1 Recruitment of researchers; Period: 2001 ongoing; Budget €12m (to date)
This programme was established to increase the number of researchers in business obtaining a PhD. The main purposes of the measure are: to rejuvenate the human capital in S&T; to employ more researchers in the business sector and then to increase business research groups formation; to link pure research with business needs that will foster innovation and research and increase competitiveness of the firm.

Poland: Support to innovation centres; Category 4.2.1 Support to innovation management and advisory services; Period: 2007–13; Budget: €190m
The main goal of the measure is to support the creation and development of innovation centres, which should be situated in the areas with great innovative potential. The innovative centres will provide the innovative entrepreneurs and researchers with complex services supporting implementation and diffusion of new technological ideas. New companies face several problems at the initial stage of their operation, most notably they suffer from the lack of institutional, technical and advisory services support. In order to facilitate the creation of new companies and successful implementation of innovative ideas, the measure will promote setting up innovation centres that will provide all necessary support for young companies, e.g. consultancy services and respective promotional activities.

Measures Addressing Institutional Failures

Measures addressing institutional failures may include any actions designed to improve the functioning or add functions to specific organizations (e.g. assisting universities to develop improved research management or industrial liaison services), creation of new institutions in an innovation system (innovation centres, patent libraries, public venture capital funds, etc.).

Hungary: Support to activities of Regional Innovation Agencies (RIAs) facilitating RTDI; Category: 4.2.1 Support to innovation management and advisory services; Period: 2007–10; Budget: €14m

This is a continuation of a scheme that has been operating since 2005. Originally the main objective was to set up so-called Regional Innovation Agencies (consortia of regional RTDI players, RIAs) which facilitate the development and strengthening of the regional innovation systems. RIAs have been established in all seven regions. Their activities are based on the regional innovation strategies and are expected to facilitate regional innovation processes, coordinate technological innovation networks, and provide innovation-related services. They operate as networks, based on partnership among interested partners. These agencies work to improve cooperation between the different organizations, coordinate funds available for innovation, generate additional funding, and promote the creation of national and international innovation networks.

Slovakia: The Fund of Funds (the former Seed Capital Company); Category: 4.3.2 Support to risk capital; Period: 1994 ongoing; Budget: €19.7m (in 2010)

The Fund of Funds (Fond fondov, s.r.o., FoF) was established by the National Agency for Development of SME in 1994. The FoF is a limited company, whose long-term mission is to properly coordinate activities of individual funds and thus stimulate the development of the sector of small and medium sized enterprises, leverage the volume of financial facilities of individual funds and use the generated profit for further support of small and medium enterprises. The scheme resources facilitated the foundation and development of many small and medium sized enterprises. By 2009 the FoF managed eight funds, established with public funds, including the INTEG Fund established in 2005 designed to support innovative projects of companies involved in the technology incubators and research-based spin-offs.

Measures Addressing Network Failures

The measures include a range of initiatives aimed at fostering greater cooperation and clustering notably amongst SMEs, user-driven innovation but also more classic industry–academic R&D cooperation programmes.

Czech Republic: COOPERATION; Category: 1.3.1 Cluster framework policies; Period: 2007–13; Budget: €190m
The programme COOPERATION implements Priority Axis 5 'Environment for Enterprise and Innovation' of the Operational Programme Enterprise and Innovation 2007–2013. This programme aims at providing support for establishing and developing cooperative sectoral associations: clusters, poles of excellence, technology platforms and cooperative projects at regional, supra-regional and international levels as a tool for boosting competitiveness of the economy and economic growth.

Measures Addressing Framework Failures

Measures addressing framework failures often concern regulatory initiatives to lighten the burden on doing business or innovating, improving IPR regimes or financial frameworks, etc. Some efforts to shift towards demand-side policies including using public procurement could be included here.

Poland: Management of intellectual property rights; Category: 5.3.2 Consultancy and financial incentives to the use of IPR; Period: 2007–13 Budget: €39m
The main goal of this measure is to improve the functioning of the innovation market and promotion of innovative solutions through the intensification and encouragement of intellectual property rights application. More specifically, the measure covers costs related to the application and protection of IP rights, especially abroad, as well as raising awareness about the importance of IP issues among entrepreneurs.

Hungary: Raising social awareness for innovation (INNOTARS_08); Period: 2008–10; Budget: €2.8m
The objective of this measure is to support a wide range of social scientific research with the aim of underpinning RTDI policies. Focus is on wider socioeconomic aspects of RTDI and policies aimed at fostering RTDI activities. More generally, the scheme supports projects aimed at offering answers to several key

challenges faced by Hungarian society. These include the ageing population, the increased role of knowledge and skills in socio-economic development etc., and these have an impact on how government policies should respond. Therefore, this scheme provides support for projects aiming to provide more detailed and useful insight into some of the key socioeconomic trends and challenges, as well as forums for debates in order to contribute to more relevant and efficient policies, especially in the field of RTDI.

Source: Adapted from TrendChart database of policy measures, selection by author.

- on average CEEC5 invest more than the EU25 on support for innovative start-ups, support for doctoral studies (funding doctoral schools and studies, and so on) and support for innovation management. Much of the support for innovative start-ups is focused on building science and technology parks, incubators and facilities, rather than on financial and advisory services to firms;
- the CEEC5 spend relatively less public money than in the EU25 on risk capital (2.2 per cent compared to 9.9 per cent), suggesting that the scope and sophistication of public–private partnerships fostering seed and other forms of equity capital for new technology based firms, may be lagging behind.

Compared to the CEEC5, in the EU25 there is a much larger share of funding allocated to three categories: support for organizational innovation (6 per cent EU25); funding for public research organizations (6.7 per cent EU25) and tax incentives for business R&D (6.6 per cent EU25). Only the Czech Republic has an allocation similar to the EU25 average for support for organizational innovation (6.6 per cent Czech Republic). The absence of funding allocated specifically to public research organizations in the CEEC5 may in part be a methodological issue (not being considered as a top priority in any measure), but also reflects the restructuring of R&D institutes that took place after 1990. Finally, tax incentives in the CEEC5 for corporate R&D, where they exist (Poland and Hungary), are reported to be largely ineffective. This may reflect the push in the 1990s towards 'neutral' taxation systems in these countries.

Other differences with the EU27 average are due to specific measures in one country, rather than being group-wide phenomena. This is the case for

Table 5.3   *Distribution of estimated public expenditure on RTDI policy measures in CEEC5 versus EU25 (2008)*

Main policy priority as percentages	EU25	CEEC5	Czech R	Hungary	Poland	Slovak R	Slovenia
4.2.3 Support for technology transfer between firms	2.4	18.4	11.1	13.5	23.6	27.1	20.1
2.3.1 Direct support for business R&D (grants and loans)	10.3	15.8	27.0	11.9	13.2	23.7	9.6
1.3.2 Horizontal measures to support financing	6.2	9.4	7.6	24.3	2.3	0.0	5.8
2.2.3 R&D cooperation projects	10.1	9.1	8.7	15.6	6.5	0.0	6.9
2.1.4 Research infrastructures	5.6	8.4	0.0	3.9	16.7	7.1	0.0
4.3.1 Support for innovative start-ups incl. gazelles	5.9	7.1	12.7	9.1	5.0	0.0	1.5
3.1.3 Stimulation of PhDs	1.2	6.4	0.0	0.4	12.4	1.2	17.0
1.2.1 Strategic research policies	5.6	4.8	0.6	0.2	6.0	3.1	30.6
1.3.1 Cluster framework policies	7.2	4.3	5.3	9.9	1.3	0.0	0.0
4.2.1 Support for innovation management and advisory services	3.1	4.1	4.5	3.2	3.3	17.9	0.7
4.1.2 Support for innovation in services	0.6	2.4	12.7	0.0	0.0	0.0	0.0
4.3.2 Support for risk capital	9.9	2.2	0.0	0.5	2.3	19.9	0.0
Sum of all other priorities	31.9	7.5	9.7	7.5	7.5	0.0	7.9

Note: Budgets for measures for the period 2007–13 are allocated to the main priority (from a maximum of four) selected for each support measure. This procedure clearly creates a 'concentration' effect but is the most neutral way of allocating the budgets in the absence of more detailed data.

Source: Author's calculations based on TrendChart-ERAWATCH data of research and innovation policy measures, data extracted June 2009.

greater relative spend on horizontal measures to improve availability of finance for companies, explained by a major SME loan guarantee measure in Hungary. Slovakia's high share of support for risk capital is due to a measure co-financed by the EU Structural Funds JEREMIE (Joint European Resources for Micro-to-Medium Enterprises) initiative, and is probably an overestimate since part of the funding is for more classic business development support and not innovation per se. However, the CEEC5 appear to spend relatively more on average than the EU25, on innovation in services, but this is related to a single funding measure (supporting ICT-related service centre investments and innovations) in the Czech Republic.

There are several notable differences within the CEEC5 group, and the Czech Republic policy mix appears to be shifting towards something more akin to the policies pursued in Western Europe, including a shift towards more service-oriented R&D and innovation measures and, along with Hungary, a stronger emphasis on R&D cooperation type measures linking the (relatively stronger) science base to business. Poland and Slovenia are the most strongly focused on boosting human resources for innovation with (in the context of each system) significant programmes promoting doctoral studies, renewal of higher education schemes, and so on. Specific measures related to mobility of researchers, recruitment of innovation staff, and so on, tend to be small-scale compared to more traditional 'broad-based' life-long learning or recruitment subsidies. Finally, Slovakia appears to be the most 'investment driven' in terms of policy mix, with high shares of funding for technology transfer, direct support to R&D and new measures to boost financial support for business development.

5.4 TOWARDS DIFFERENTIATED INNOVATION POLICIES: LESSONS FOR THE CEEC

What conclusions can be drawn from the available evidence on how well innovation policies in the CEEC5 are tailored to meeting the specific challenges (or failures) of NSI? This chapter has argued that the CEEC5 have specific strengths and weaknesses in their innovation performance that require them to develop tailored policies and avoid blindly following a 'Europeanized' policy approach (Radosevic and Reid, 2006). While the policy objectives set out in government strategies are broadly in line with the innovation challenges facing the CEEC5, closer analysis of the policy mix would suggest that there is a need for further refinement of policy measures and methods.

Since 2004 there has been a significant increase in public budgets for STI, which has resulted in a large number of new support measures in the CEEC5. As Tsipouri and Reid (2010) underline, based on the 2009 TrendChart reports, in the last five years, in most of the EU27 member states, there has been an increase in STI funding and, in some cases, such as Slovakia, it has been a four- to fivefold increase annually since 2004. This increase is in part due to the Structural Funds effect on new member states since 2004, but 'leader-followers' (using EIS groupings) have also received significant increases. However, the effects of the global financial crisis are threatening to re-open the funding gap between the leaders and the catching-up countries, which was beginning to close up. The TrendChart reports for 2009 highlight increased budgets in leading EU27 countries (for example, Finland, Sweden) in response to the crisis, and budget cuts in moderate and catching-up countries. An expected stabilization effect of the multi-annual Structural Funds programming effort does not appear to have protected (R&D and) innovation funding from cuts, in the less advanced member states.

In terms of the breadth of innovation policy, the CEEC5 tend to focus on direct financial intervention for applied R&D and development of research and innovation infrastructures (science parks, and so on). The budgetary snapshot provided in this chapter confirms that the CEEC5 policy mix differs from the 'average' EU25 mix and shows that in a number of cases the emphasis/focus of government intervention is on resolving specific issues related to the stage of innovation (for example investment-driven approach to RTDI policy via technology acquisition by SME or promotion of research infrastructure), tackling specific bottlenecks in NSI (for example doctoral studies in science and engineering, low capabilities and spend of domestic enterprises on R&D), and so on.

However, at the time of writing (spring 2010), there is little evidence that such an innovation policy mix is being effective in boosting innovation activity (witness the limited convergence in innovation peformance). This may be a reflection of insufficient attention to framework and institutional failures and lack of policy measures targeting the demand side and aimed at overcoming the 'socio-cultural' barriers to innovation, which are seen as a critical weakness in the NSI of the CEEC5. Policy measures to increase public understanding of the role of science and technology and the risks and rewards related to innovation are important. Support for non-technological innovation (boosting design and creativity) is not sufficiently high up the policy agenda, and, while human resources for innovation are both a policy challenge and a policy priority, in operational terms the funding directed to innovation management-related skills (as opposed to general employee training schemes) remains limited (for example in Slovenia; see Bucar, 2008).

The approach to innovation policy in the CEEC5 is largely focused on industrial or technological development as a form of updated industrial policy. This might be justified by the state of economic development in these five countries, but evidence from statistical analyses and policy studies, such as the TrendChart, suggests that there is a need to widen the scope of innovation policy to take account of linkages with other policies (such as education and training, in line with Aghion's (2006) arguments) rather than simply trying to remedy the perceived underinvestment in business R&D by supporting what Metcalfe (2005) described as 'individual innovation events' (through direct subsidies for applied research or to support the purchase of embodied technology). In view of the high share of equipment in the innovation expenditure of the CEEC5 firms, this will need to be complemented by 'soft' investments (training, and techniques such as design, and other skills in the workforce) in line with the high share of non-R&D innovators in these economies.

ACKNOWLEDGEMENTS

I am grateful to Professor Slavo Radosevic for comments on an earlier draft of this chapter. The chapter is based on work undertaken in 2007–09 in the framework of the project INNO-POLICY Trend Chart (contract number 046381), funded by the EC through the 6th Framework Programme. Numerous discussions, ideas and contributions during the lifetime of the projects are gratefully acknowledged from the members of the project team, notably Michal Miedzinski, Miriam Ruiz Yaniz, Professor Lena Tsipouri and Dr Paul Cunningham. As always, any remaining errors are those of the author.

NOTES

1. Cowan and Van de Paal (2000) define innovation policy as a set of policy actions to raise the quantity and efficiency of innovation activities whereby innovation activities refer to the creation, adaptation and adoption of new or improved products, processes or services. Similarly, Lundvall and Borrás (2005) make a distinction between science policy, concerned with the production of knowledge, and innovation policy, which aims to improve the overall innovative performance of the economy.
2. In the literature on NSI, the term 'institution' is taken to mean rules and routines such as the legal system, informal rules and their enforcement characteristics. Here, as in everyday practice, the term 'institution' is used as a synonym for 'organization'.
3. The policy mix concept is based on the idea that it is the combination of policy

instruments interacting with each other that influences R&D and innovation performance rather than instruments taken in isolation. The second key idea is that R&D and innovation are not only influenced by 'innovation policy' but also by other policies such as, for example, environmental regulations, regulation of energy systems, and so on. See Nauwelaers (2009) for a more in-depth review.

4. However, the current NACE code industry classification system is becoming inappropriate as a mechanism for describing a set of common business activities, especially in the 'traditional sectors' (e.g. technical textiles, apparel, fine chemicals, supplier of industrial materials, supplier to the automobile industry). If there are differences in the sectoral innovation system they may be hard to capture due to the multi-sector activities of many innovative and especially large firms. See the body of evidence on sectoral innovation systems developed under the EU-funded Sectoral Innovation Watch project (2005–2008) (Reinstaller and Unterlass, 2008).

5. The industrial structure of the EU27 is relatively more concentrated in medium-high-tech, medium-low-tech and low-tech activities than in the US, which is more concentrated in high-tech sectors, notably information and communication technology (ICT) manufacturing industries and air and spacecraft.

6. Tödtling and Trippl (2005) make a similar argument about a lack of any 'ideal model' of innovation policy, due to regional innovation systems exhibiting very different barriers to innovation.

7. The reflection can be extended by considering relative R&D productivity and specialization of a limited number of EU27 regions which dominate in terms of R&D expenditure and outputs. See Peter and Frietsch (2009).

8. Gault and Huttner (2008) argue that it is difficult for public policy to influence complex systems and that a 'whole-government' approach requires coherent and comprehensive analysis of system dynamics, and robust statistical measures that describe the actors and processes at work. They give the example of R&D subsidies: it might be expected that high subsidies would enhance the flow of knowledge, but their impact might be eroded by relatively high corporate taxes, necessary to fund the subsidies.

9. The methodology for calculating the summary score for each group of EIS indicators and the overall summary innovation index is explained at: http://www.proinno-europe. eu/page/technical-annex-choice-indicators-and-methodology.

10. R&D innovators are defined as all innovators performing in-house or intramural R&D. Non-R&D innovators innovate by acquiring or by buying extramural R&D (that is R&D performed by other companies or research organizations), by buying advanced machinery, equipment and computer hardware or software, by buying or licensing patents and non-patented inventions, by training their personnel, or by spending resources on the design or market introduction of new goods or services.

11. Each TrendChart country correspondent (national expert) is asked to identify, on an annual basis, the three key challenges for innovation policy, based on an analysis of EIS findings for their country and any additional complementary national studies or reports.

12. http://www.proinno-europe.eu

13. http://cordis.europa.eu/erawatch/

14. A single support measure can be assigned up to four policy priorities in the database. Percentages refer to the share of support measures addressing a given policy priority for each EIS country group. The numbers in brackets in the key indicate the total number of measures for each EIS group. The chart includes 12 out of 40 policy categories resulting from the fusion of two sets: the 10 policy priorities addressed by the highest number of support measures and the 10 priorities with the highest variance between the smallest and biggest share of measures for EIS groups.

15. For a full methodological explanation, see Tsipouri and Reid (2010).

REFERENCES

ADE (2001), 'Innovation policy in six candidate countries: the challenges', *Innovation Papers no. 16*, Brussels: European Commission, available at http://www.proinno-europe.eu/node/20590, accessed 24 February 2010.

ADE (2003), 'Innovation policy in seven candidate countries: the challenges (Volume 1)', *Innovation Papers no. 34*, Brussels: European Commission, available at http://www.proinno-europe.eu/node/20591, accessed 24 February 2010.

Aghion, P. (2006), 'A primer on innovation and growth', *Bruegel Policy Brief, 2006/06*, available at www.bruegel.org/uploads/tx_btbbreugel/pbf_061006_innovation.pdf, accessed 24 February 2010.

Aho, E. et al. (2006), 'Creating an innovative Europe, report of the independent expert group on R&D and innovation', available at http://ec.europa.eu/invest-in-research/pdf/download_en/aho_report.pdf, accessed 24 February 2010.

Arnold, E. (2004), 'Evaluating research and innovation policy: a systems world needs systems evaluations', *Research Evaluation*, **13**(1), 3–17.

Arrow, K.J. (1962), 'Economic welfare and the allocation of resources for invention', in R.R. Nelson (ed.), *The Rate and Direction of Inventive Activity*, Princeton, NJ: Princeton University Press, pp. 609–25.

Arundel, A., C. Bordoy and M. Kanerva (2008), 'Neglected innovators: how do innovative firms that do not perform R&D innovate? Results of an analysis of the Innobarometer 2007 survey No. 215', INNO-Metrics Thematic Paper, available at http://www.proinno-europe.eu/page/admin/uploaded_documents/EIS%202007%20Neglected%20innovators.pdf, accessed 23 June 2010.

Balasz, V. (2008), 'Annual Inno-Policy TrendChart report for Slovakia', available at http://www.proinno-europe.eu/node/20531, accessed 24 February 2010.

Bruno, N., M. Miedzinski, A. Reid and M. Ruiz Yaniz (2008), 'Socio-cultural determinants of innovation', Brussels: Technopolis Group, available at http://www.technopolis-group.com/resources/downloads/Socio-cultural-factors-innovation.pdf, accessed 23 June 2010.

Bucar, M. (2008), 'Annual Inno-Policy TrendChart report for Slovenia', available at http://www.proinno-europe.eu/node/20529, accessed 24 February 2010.

Cowan, R. and G. van de Paal (2000), *Innovation Policy in a Knowledge-based Economy*, Brussels: European Commission, Directorate-General for Enterprise.

Duchêne, V., E. Lykogianni and A. Verbeek (2009), 'The EU R&D under-investment: patterns in R&D expenditure and financing', in H. Delanghe, U. Muldur and L. Soete (eds), *European Science and Technology Policy: Towards Integration or Fragmentation?*, Cheltenham, UK and Northampton, MA, USA: Edward Elgar, pp. 193–213.

Edler, J. (2005), 'Innovation and public procurement: review of issues at stake', Study for the European Commission ENTR/03/04, available at http://www.proinno-europe.eu/node/20569, accessed 23 June 2010.

European Commission (2007), *European Innovation Scoreboard*, Luxembourg: EC.

Freeman, C. (2006), 'Catching up and innovation systems: implications for Eastern Europe', in K. Piech and S. Radosevic (eds), *The Knowledge based Economy in Central and Eastern Europe: Countries and Industries in a Process of Change*, London: Palgrave Macmillan, pp. 13–30.

Gault, F. and S. Huttner (2008), 'A cat's cradle for policy', *Nature*, **455**, 462–3.

Georghiou, L. (2007), 'Demanding innovation: lead markets, public procurement and innovation', London: NESTA, available at http://www.nesta.org.uk/assets/documents/demanding_innovation, accessed 14 February 2010.

Havas, A. (2008), 'Annual Inno-Policy TrendChart report for Hungary', available at http://www.proinno-europe.eu/node/20530, accessed 14 February 2010.

Knell, M. (2008), *Product-Embodied Technological Diffusion and Intersectoral Linkages in Europe. Europe Innova Sectoral Innovation Watch report*, Brussels: European Commission, DG Enterprise, available at http://www.europe-innova.eu/c/document_library/get_file?folderId=24913&name=DLFE-2646.pdf, accessed 24 June 2010.

Lin, J.Y. (2010), 'New structural economics: a framework for rethinking development', *World Bank, Policy Research Working Paper 5197*, Washington, DC: World Bank.

Louis Legrand & Associés, PREST (University of Manchester) and ANRT (2002), *Innovation Tomorrow: Innovation policy and the regulatory framework: making innovation an integral part of the broader structural agenda*, European Commission, Brussels, available at http://www.proinno-europe.eu/node/20564, accessed 23 June 2010.

Lundvall, B.-Å. and S. Borrás (2005), 'Science, Technology and Innovation Policy', in J. Fagerberg, D.C. Mowery and R. Nelson (eds), *The Oxford Handbook of Innovation*, Oxford: Oxford University Press, pp. 599–631.

Malerba, F. (2005), 'Sectoral systems: how and why innovation differs across sectors', in J. Fagerberg, D.C. Mowery and R. Nelson (eds), *The Oxford Handbook of Innovation*, Oxford: Oxford University Press, pp. 380–406.

MERIT (2007), 'European Innovation Scoreboard 2007', available at http://www.proinno-europe.eu/page/admin/uploaded_documents/European_Innovation_Scoreboard_2007.pdf, accessed 26 February 2010.

MERIT (2009), 'European Innovation Scoreboard 2008', available at http://www.proinno-europe.eu/page/admin/uploaded_documents/EIS2008_Final_report-pv.pdf, accessed 26 February 2010.

MERIT (2010), 'European Innovation Scoreboard 2009', available at http://www.proinno-europe.eu/sites/default/files/page/10/03/I981-DG%20ENTR-Report%20EIS.pdf, accessed 26 February 2010.

Metcalfe, J.S. (2005), 'Systems failure and the case for innovation policy', in P. Llerena and M. Matt (eds), *Innovation Policy in a Knowledge Based Economy*, Berlin: Springer, pp. 47–74.

Nauwelaers, C. (2009), 'Final report of the policy mix study', Brussels: European Commission, Directorate-General Research, available at http://www.policymix.eu/policymixtool/page.cfm?pageid=322, accessed 23 June 2010.

Organisation for Economic Cooperation and Development (OECD) (1997), 'National Innovation Systems', available at http://www.oecd.org/dataoecd/35/56/2101733.pdf, accessed 26 February 2010.

Organisation for Economic Cooperation and Development (OECD) (2005), *The Oslo Manual: Guidelines for Collecting and Interpreting Innovation Data (3rd edn)*, Paris: OECD.

Pazour, M. (2008), 'Annual Inno-Policy TrendChart report for Czech Republic', available at http://www.proinno-europe.eu/node/20532, accessed 26 February 2010.

Peter, V. and R. Frietsch (2009), 'Exploring regional technology specialisation: implications for policy', *Regional Key Figures of the ERA Series*, Brussels:

European Commission, DG Research, available at http://ec.europa.eu/invest-in-research/pdf/download_en/kina24049enn.pdf, accessed 26 February 2010.

Porter, M. (1990), *The Competitive Advantage of Nations*, New York: The Free Press.

Radosevic, S. (2006), 'The knowledge based economy in central and Eastern Europe: an overview of key issues', in K. Piech and S. Radosevic (eds), *The Knowledge Based Economy in Central and Eastern Europe: Countries and Industries in a Process of Change*, London: Macmillan, pp. 31–53.

Radosevic, S. and A. Reid (2006), 'Innovation policy for a knowledge based economy in Central and Eastern Europe: driver of growth or new layer of bureaucracy?', in K. Piech and S. Radosevic (eds), *The Knowledge Based Economy in Central and Eastern Europe: Countries and Industries in a Process of Change*, London: Macmillan, pp. 295–311.

Reinstaller, A. and F. Unterlass (2008), 'What is the right strategy for more innovation in Europe? Drivers and challenges for innovation performance at the sector level', Final report of the Sectoral Innovation Watch project, Brussels: European Commission, DG Enterprise, available at http://www.europe-innova.eu/c/document_library/get_file?folderId=24913&name=DLFE-2674.pdf, accessed 24 June 2010.

Rodrik, D. (2004), 'Industrial policy for the twenty-first century', mimeo, John F. Kennedy School of Government, Harvard University, Cambridge, MA.

Rostow, W.W. (1960), *The Stages of Economic Growth, A Non-Communist Manifesto*, Cambridge: Cambridge University Press.

Smith, K. (2000), 'Innovation as a systemic phenomenon: rethinking the role of policy', *Enterprise and Innovation Management Studies*, **1**(1), 73–102.

Technopolis (2006), 'Strategic evaluation of innovation and knowledge in the structural funds', Brussels: European Commission, Directorate-General for Regional Policy, available at http://ec.europa.eu/regional_policy/sources/docgener/evaluation/pdf/strategic_innov.pdf, accessed 24 June 2010.

Tödtling, F. and M. Trippl (2005), 'One size fits all? Towards a differentiated regional innovation policy approach', *Research Policy*, **34**(8), 1203–19.

Tsipouri, L. and A. Reid (2010), 'European innovation progress report 2009', Brussels: European Commission, available at http://www.proinno-europe.eu/sites/default/files/page/10/05/EIPR2009%20final.pdf, accessed 26 February 2010.

Tsipouri, L., A. Reid and M. Miedzinski (2009), *European Innovation Progress Report 2008*, Brussels: European Commission, DG Enterprise, available at http://www.proinno-europe.eu/node/admin/uploaded_documents/EIRP2008_Final_merged.pdf, accessed 26 February 2010.

von Hippel, E. (1988), *The Sources of Innovation*, Oxford: Oxford University Press.

Walendowski, J. (2008), 'Annual Inno-Policy TrendChart report for Poland', available at http://www.proinno-europe.eu/node/20533, accessed 26 February 2010.

World Economic Forum (WEF) (2009), *The Global Competitiveness Report 2009–2010*, available at http://www.weforum.org/pdf/GCR09/GCR20092010fullreport.pdf, accessed 26 February 2010.

6. Attracting and embedding R&D in multinational firms: policy options for EU new member states

Rajneesh Narula

6.1 INTRODUCTION

European Union (EU) integration and expansion is a complex, cooperative, socioeconomic undertaking. At the most basic level, it requires the political, economic and sociological milieux of new member states (NMS) to converge with those of the core member state countries. Each new wave of EU members results in another multiple group of countries, at a different stage of development, with different levels of gross domestic product (GDP), resource endowments, comparative advantages and industry and economic structures. This diversity also results in different growth trajectories.

It is increasingly obvious that continual EU expansion means that a common set of industrial and technology policies to promote growth, applicable across all member states, or even a single set of targets, is not feasible. In this chapter, we view the EU as consisting of groups of countries that have some common features, which can be the basis for some common policy recommendations. We group the NMS into two sub-groups, which reflect the extent of their convergence and development of a market economy. The first group includes the Czech Republic, Slovakia, Hungary, Slovenia, Estonia and Poland, which we refer to as the advanced NMS. The second group includes Bulgaria, Romania, Latvia and Lithuania, which are still in a stage of transition. In this chapter we refer to them collectively as the new NMS.[1]

The chapter focuses on the promotion of innovation activities by multinational enterprises (MNE) in the NMS. This requires a somewhat different approach than promotion of general value-adding activities by MNEs. At the beginning of the 1990s, the operations of MNEs tended to be miniature replicas of the home country operations, with most or all aspects of the value chain being undertaken in each host country.

This set-up now applies only to the case of very few investments, reflecting important changes in the global economic context, associated with increasing interdependencies between countries, industries and firms. MNEs are distributing their activities across regions and countries in order to exploit the technological capabilities of locations most efficiently in a way that best suits specific aspects of the activities. MNEs are also increasingly fragmenting their innovative activities across different locations to exploit specific aspects of particular systems. In certain cases, these may be demand-oriented – such as the presence of large markets or the availability of generic price-sensitive inputs. These are the centrifugal factors that promote the establishment of production and other value-adding activities of MNEs as they attempt to exploit their existing assets and competences in conjunction with locally sourced inputs. Such innovation activities are a case of research and development (R&D) responding to demand conditions. In other words, innovation consists of adaptations to existing products and services to suit local demand. The R&D tends to be low knowledge-intensive and to remain somewhat independent, requiring greater integration with the parent firm than with local knowledge infrastructure. It requires MNE market-seeking foreign direct investment (FDI) activities to integrate forward into R&D. We refer to this as demand-driven R&D.

In other circumstances, MNEs may choose to establish themselves in particular locations especially (in some cases only) to undertake innovation because of specific, location-bound assets, which may or may not include quasi-public goods provided through universities and public research institutes. These types of innovation activities involve stand-alone R&D facilities that are considerably more knowledge-intensive than in the case of demand-driven R&D, and imply considerably greater dependence on domestic knowledge sources and infrastructures. We refer to this as supply-side R&D.

These two types of R&D require somewhat different approaches and, necessarily, different policy options. Technology and industrial policies are inextricably linked to the outcomes of FDI policies. Section 6.2 discusses and develops the innovation systems framework as a basis for understanding the effects of the interactions between technological capabilities and quasi-public goods. Sections 6.3 and 6.4 discuss the complexities introduced by globalization and extend the innovation systems approach to allow for cross-border influences and relationships, and the interlocking nature of FDI, and industrial and innovation policies. Section 6.5 provides some suggestions for NMS governments about how to link FDI and innovation policy.

6.2 THE INNOVATION SYSTEMS AS A BASIS FOR ANALYSING POLICY OPTIONS

This chapter begins from the premise that all economic actors expand their activities depending upon the strength (or weakness) of their competitive assets. These assets include technological assets, in the sense of ownership of plant, equipment and the technical knowledge embodied in their engineers and scientists. In addition, economic units of all sizes possess competitive advantages that derive from: (a) the ability (i.e. knowledge) to create efficient internal hierarchies (or internal markets) within the boundaries of the firm; and (b) the ability to utilize external markets efficiently.

Economic actors refer to organizations that are engaged in the regular production of outputs (whether physical goods or services) to meet a specific or general demand. Economic actors fall into two groups. The first group are firms – private and public – engaged in innovative activity. Although they may not always be organized with the primary intention of generating economic rents (as is the case for state-owned firms), ongoing activities are evaluated on the basis of achieving their owner-defined output criteria. The second consists of non-firms that determine the knowledge infrastructure that supplements and supports firm-specific innovation, whose objective may be to make available their outputs as semi-public goods. Economic actors are distinguishable from political and social actors. These political and social actors do not generate innovative outputs per se, but their actions and activities shape the nature of the activities of the economic actors.

While innovation may take place at firm-level, firms exist as part of a system. They are embedded in these systems through historical, social and economic ties to other economic and non-economic actors. Thus in order, from a policy perspective, to understand innovation (or the lack thereof) we need to understand the systemic interactions, relationships and routines of all the organizations involved and their complex interactions with the firm. The firm's environment is made up of interactions between firms – and especially between a firm and its network of customers and suppliers. The environment also encompasses factors that shape the behaviour of firms: the social and cultural context; the institutional and organizational framework; the infrastructures; the processes that create and distribute scientific knowledge, and so on.

Figure 6.1 provides a stylized version of a 'conventional' national system of innovation (NSI). By conventional, we mean an innovation system that is typical of a non-socialist, market economy and that, essentially, is representative of the EU core member countries. A systems of

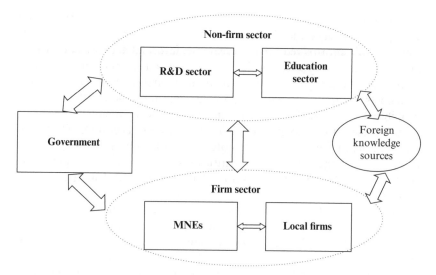

Source: Narula and Jormanainen (2008: 34).

Figure 6.1 The conventional model of an innovation system

innovation approach allows us to map the complex interactions between the firm and its environment.

In addition to the firm and non-firm sectors, which account for most innovation activity, government works to create and disseminate, acquire and utilize knowledge. In this chapter we conceive government generically, and include governments motivated internally (at country level) and externally (at supra-national level).

The interactions between the various actors within a system are governed by institutions. Institutions can be of two types – informal and formal – and generally are understood as 'sets of common habits, routines, established practices, rules, or laws that regulate the interaction between individuals and groups' (Edquist and Johnson, 1997: 43). We understand formal institutions as the intellectual property rights (IPR) regime, competition policy, technical standards bodies, taxation regime, the incentives and subsidies for innovation, funding of education, and so on. Formal institutions generally involve politically defined and legally binding rules, regulations and organizations, and the political and economic spheres are rarely independent, especially when they involve high levels of central planning. In general, the policy environment in which economic actors function has a high degree of interdependence with the economic and political spheres.

Developing and modifying informal institutions involves complex and slow processes since they cannot be created simply by government fiat. Considerable effort is required to create the informal networks of government agencies, suppliers, politicians and researchers and, once created, there is a low marginal cost for their maintenance. For outsiders, the high costs of familiarity with, and integration into a new system may be prohibitive (Narula, 2003). For the insiders, however, membership comes with privileges that provide opportunities for rent generation. Studies of informal institutions – which are notoriously difficult to quantify – point to the absence or inefficiency of institutions as a primary force inhibiting economic development (for example Rodrik, 1999; Rodrik et al., 2004; Asiedu, 2006).

The former centrally planned economies among the NMS had national and rather closed economic systems, and a fundamentally different structure. Figure 6.2 depicts a stylized version of a pre-transition system of innovation model. Prior to economic reform, the knowledge sources in the innovation systems of transition economies were determined primarily by domestic elements (Radosevic, 1999, 2003). Their technological development trajectories were planned centrally in response to state-defined priorities. Likewise, domestic government organizations formulated domestic industrial policy, which, in turn, determined the domestic industrial structure. National, non-firm actors defined the types of skills required, the kinds of technologies in which the local labour force was required to acquire expertise; the kinds of technologies in which basic and applied research was conducted and, thus, the industrial specialization and competitive advantage of their economies' firm sectors. FDI was non-existent prior to transition, and any linkages to international sources were sporadic and state-controlled.

One of the primary conditions for EU membership is that candidate countries' economic systems must demonstrate convergence towards the EU norm, which necessarily has meant that NMS from the former centrally planned economies needed to demonstrate that they are functioning market economies.

Although some NMS countries have responded successfully to radical changes to their industrial structures, the response of others has been less successful, and this broadly reflects the division between 'advanced NMS' and 'new NMS'. The two groups have fundamentally different policy regimes, with some persisting with domestic firm-led industrialization and others moving to an MNE-led development strategy (Radosevic, 2006). Broadly speaking, the latter group have modified their institutions and attempted to redesign their innovation systems around the 'conventional' market economy model, with varying degrees of success (Radosevic,

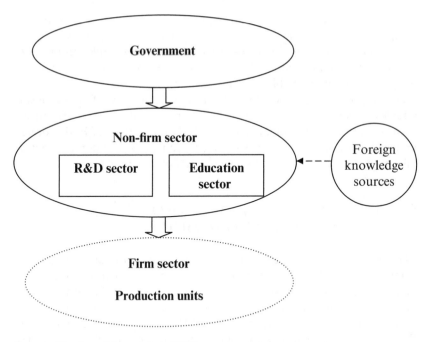

Source: Narula and Jormanainen (2008).

Figure 6.2   A pre-transition model of innovation systems in centrally planned countries

2006). For the most part, their success depends on the existence of a political imperative and a popular sentiment to establish a distance from pre-transition dependence on the Soviet Union, and to realign with the EU. In other words, the ability of different economies to achieve successful transition reflects the strength of the existing institutional arrangements and the political will to implement reforms (Newman, 2000).

There are two levels of government at work in all EU economies: the formal institutions established by national governments, and those promulgated by EU law, and implemented by the European Commission (EC). Bruszt and McDermott (2008) argue that the supranational institution-building required by the EU impinges greatly on the restructuring of national institutions in post-communist states. It alters the supply side by making resources available to states to overcome the inertia of interest groups and other entrenched actors in the face of change while, at the same time, affecting the demand for institutional change by empowering actors to participate in institution-building through the creation of linkages between domestic and foreign actors. EU-wide regulatory and competition policy, and social and economic treaties and the like are binding and override national law. As we discuss later, while this provides certain location advantages relative to non-member states, it constrains the policy options available to member states.

It is important also to emphasize that the significance of non-domestic knowledge sources has changed quite dramatically, and not entirely because of the growing significance of MNE affiliates, or of FDI. Foreign knowledge sources and associated interdependencies with domestic actors take many forms, which are discussed in detail in section 6.3.

6.3   THE ROLE OF FOREIGN KNOWLEDGE IN AN INNOVATION SYSTEM MODEL

The sources of knowledge available in a typical national system are a complex blend of domestic and foreign sources, as illustrated in the simplified (stylized) framework depicted in Figure 6.3. Learning processes are not limited to intra-national interaction; increasingly they include international interaction. The pervasive role of MNE in a globalizing world, and their ability to utilize technological resources located elsewhere, makes the use of a purely NSI approach rather limiting.

In domestic innovation systems (such as pre-transition Central and Eastern Europe (CEE) NMS), the path of technological development was determined primarily by domestic elements. Technological development is now driven largely by the changing demand from local customers, which, pre-transition, was driven by government targets. Likewise, domestic government organizations determine domestic industrial policy, which, in turn, determines the industry structure. National, non-firm sources of knowledge and national universities determine the skills of a country's engineers and scientists, their technological expertise and the technological fields in which basic and applied research are conducted. They determine, therefore, national industrial specialization and the competitive advantages of domestic firms. At the end of the first decade of the twenty-first century, few such domestic systems still exist, and in most EU countries excepting the NMS, MNE subsidiaries are so well embedded that they are regarded as being part of the domestic environment. Nonetheless, the interaction between the domestic and foreign-owned firm sector varies considerably, either because domestic firms are largely in different industrial sectors, or because the two have evolved separately. In the economies of the NMS, the presence of foreign MNEs was a new phenomenon in the early post-transition years and local actors were often reluctant to

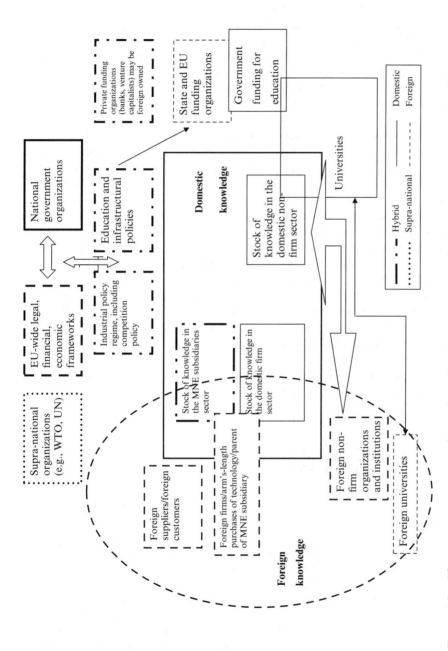

Figure 6.3   The growing non-national interdependencies on shaping knowledge and innovation in the EU

integrate MNEs into their domestic systems (Damijan et al., 2003; Sinani and Meyer, 2004; Javorcik and Spatareanu, 2008).

Universities and research institutes and their research groups collaborate with universities and research institutes in other countries.[2] The EC Framework Programmes have played an important role in facilitating cross-border collaboration between economic actors in the EU (Narula, 2003), including public research organizations and firms (Arundel and Geuna, 2004; Fontana et al., 2006).

Figure 6.3 depicts some of the actors and arenas not directly associated with knowledge creation, but which, nonetheless, are crucial for determining the efficiency with which knowledge is created, diffused and utilized in the innovation system. Perhaps most significant is the blend of EU-level organizations and national governments. EU law supersedes national regulatory frameworks, and competition policy among other forms of regulation is determined at EU level. As a precondition for EU membership countries are required to accept the *acquis communautaire*, which makes discrimination between domestic and foreign firms no longer possible. Thus, it constrains and predetermines the use of EU and domestic funds. At both global and supranational institutional levels, countries are constrained by the international treaties to which the EU or the individual country subscribes. This includes World Trade Organization (WTO) agreements such as Trade Related Aspects of Intellectual Property Rights and Trade Related Investment Measures, which shape the policy tools available to countries.

It is not the intention in this chapter to analyse the broader implications of globalization for knowledge systems in the EU. The focus remains the role of the MNE in innovation systems.

6.4 HOW DOES MNE ACTIVITY BENEFIT NATIONAL SYSTEMS?

In this section we consider the perspective of the MNE and the linkages it facilitates. This allows us to account for a variety of forms of interdependencies between and among firms, regions, countries and industries.

Linkages can be domestic (knowledge flows between the affiliate and other actors in the domestic economy) or involve foreign sources of knowledge. Figure 6.4 depicts a two-country scenario based around a joint venture (JV) between an MNE and a domestic firm. If we rely on foreign direct investment (instead of MNEs) as the unit of analysis, we limit discussion of the potential spillovers from linkages to the organizations linked by the large block arrows, which involve equity relationships.

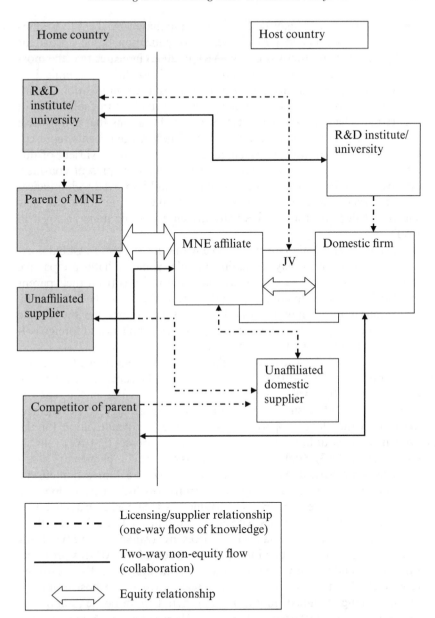

Figure 6.4   Equity and non-equity cross-border knowledge flows

However, technology may be licensed or purchased by the MNE affiliate from an unaffiliated public research organization based abroad or locally. A report by Innovation.bg (ARCF, 2006) indicates that the most innovative Bulgarian companies are interested primarily in selling their innovations to foreign firms rather than using these innovations themselves (ibid.). A second set of linkages is indicated by the solid lines in Figure 6.4. These are the active two-way collaborations that can involve several actors, both domestic and foreign. These arrangements represent higher levels of knowledge exchange, and may involve a variety of different partners. In general, these non-equity linkages present considerable potential for increased knowledge flows and improved technological competitiveness among domestic firms since they create important new sources of demand for commercially driven economic units engaged in R&D.

The nature of the affiliate and its role within its company's global portfolio of affiliates also plays a significant role. Some affiliates are passive in the sense that they receive ready-made innovations from their parent firms and, therefore, do not establish the types of linkages that enhance the indigenous innovation milieu. In other words, at one extreme, the affiliate may be operating in an enclave, utilizing foreign suppliers and foreign collaborators identified by the parent firm.

The quality of the knowledge spillovers from an investment is associated with the scope and level of competence of the subsidiary, which are co-determined by several factors (see Figure 6.5). These include factors internal to the MNE, such as internationalization strategy, role of the new location in the global portfolio of subsidiaries, and motivation for the investment, in addition to the location-specific resources that are available (Benito et al., 2003). High competence levels require complementary assets that are non-generic and are often associated with agglomeration effects, clusters and the presence of highly specialized skills. Where to locate a technology-intensive industry is constrained by local resource availability. For instance, R&D activities tend to be concentrated in a few locations where the appropriate specialized resources are available. The embeddedness of firms is often a function of the duration of the MNE's presence, since firms tend to grow their activities incrementally. MNEs most often rely on the location advantages that already exist in the host economy, and deepening of embeddedness occurs generally in response to improvements in domestic technological capacity. However, while the scope of the activities undertaken by a subsidiary can be modified fairly quickly, development of competence levels takes time. MNE investments in high value-added activities (often associated with high competence levels) tend to be 'sticky'. Firms demonstrate greater inertia when it comes to relocating

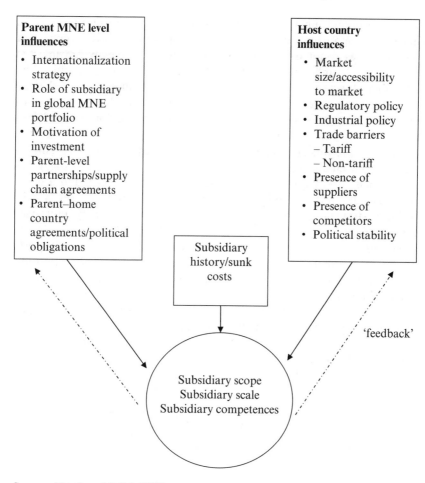

Source: Narula and Bellak (2009).

Figure 6.5 Determinants of competence, scope and scale of foreign affiliate

R&D activities. This reflects the high costs and considerable time required to develop linkages with the innovation system (Narula, 2002).

Increasingly firms are engaging in rationalizing their activities globally, so as to maximize the link to specific value-adding activities and locations with specific competitive and comparative advantages. This has led to a tendency amongst MNEs to 'break up' their value chains and locate specific aspects in particular locations to achieve maximum efficiency. It is rare for a single location to host the whole of the value chain for one product.

Prior to economic liberalization and EU integration, MNEs responded to investment opportunities primarily by establishing truncated replicas of their facilities at home, although the extent to which they were truncated varied considerably among countries. The truncation was determined by a number of factors, by far the most important determinant – and thereby the scope of activities and competence level of the subsidiary – being associated with market size, capacity and capability of the domestic industry (Dunning and Narula, 2004).

One of the results of globalization and the subsequent spatial redistribution of value chains has been that many countries have seen a downgrading of their subsidiaries in terms of scope and competence, and a move towards sales and marketing operations. Only a few have seen a shift towards strategic centres or have maintained multi-activity units (see Figure 6.6).

The use by firms of global production networks has generally been to the benefit of the MNE, while host countries with generic location advantages have seen a decline in scale, scope and competence of local value-adding activities. Competition among locations to host activities is considerable, although few can provide specialized and well-developed innovation systems. The benefit to be derived from the location of a subsidiary varies considerably. A sales office or an assembly unit may produce high turnover and employ large numbers of staff, but the technological spillovers will be relatively less than from a manufacturing facility.

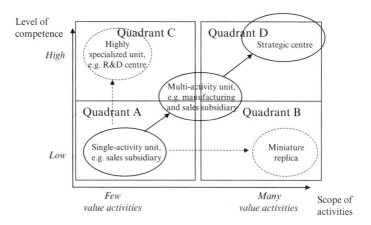

Source: Benito et al. (2003).

Figure 6.6   Types of subsidiaries and relationships to scope and competence levels

Activities such as sales and marketing and natural resources extraction are usually located in countries in the early stages of transition (and farthest away from convergence with the EU norm), which have limited domestic sectors and poorly defined innovation systems. It is in the most advanced economies with good domestic technological capacity (core EU members) that the least truncated subsidiaries (which often include R&D departments) are located (see Majcen et al., 2009 for a more in-depth discussion).

Few MNEs are continuing with the strategy of miniature replicas when engaging in greenfield investments. Rationalization of activities within the single market, in many cases, has led to the downgrading of activities from truncated replicas to single activity affiliates. MNEs have taken advantage of the EU single market to reduce the number of production locations in order to exploit economies of scale at plant level, especially where local consumption patterns are not sufficiently different to justify local capacity and where transportation costs are not prohibitive. This has resulted in some replica companies being downgraded to sales and marketing affiliates, which offer fewer opportunities for spillovers.

To what extent the NMS are and will be able to benefit from technological spillovers from MNE activity is unclear. Also, as already mentioned, although there will be some investment in new affiliates resulting in new (greenfield) subsidiaries, there will also be some downgrading of subsidiaries. MNEs may discontinue operations in certain locations in order to take advantage of the benefits to be derived elsewhere in the EU (Spain and Portugal have seen their low-cost advantages being eroded), or may reduce the intensity of their operations by reducing the scope and intensity of subsidiaries. There may also be a redistribution effect, with sectors that once were dominated by domestic capital being transferred to foreign ownership, particularly where domestic capitalists have failed to improve competitive advantage to compete effectively with foreign firms. In many NMS, the share of foreign ownership in total capital stock is already typically much higher than in the older EU member states, although with considerable variation across sectors (Narula and Bellak, 2009). Garmel et al. (2008) predict that 75 per cent of capital in the NMS will be acquired ultimately by investors from the core member states. Government incentives and subsidies are rarely pivotal in determining the scope and competence of MNEs and the potential for greater technological spillovers. We would emphasize that from a growth and learning perspective, externalities are important only if they can be captured by other economic actors in the host economy. For their optimal utilization, there needs to be an appropriate match between the nature of potential externalities and the absorptive capacities of domestic firms.

6.5   LINKING MNE POLICIES AND INNOVATION POLICIES

It is not recommended that in proposing policy the NMS should be treated as a single group: there is considerable diversity in their economic structures. There are also path dependencies that reflect each individual member country's socio-political and economic histories. In particular, member states that have completed the transition from a socialist, non-market system retain significant artefacts of their pre-transition eras in their innovation systems although, again, the extent varies among countries. Also, the roles of MNEs and other international economic actors in these countries can be fundamentally different. On the other hand, transition has produced some convergence in economic and structural policies and trajectories, which is a requirement for EU membership, and NMS policies towards MNEs have broadly converged. However, levels of institutional convergence vary, with the Czech Republic probably closest to the level in the EU core countries, and Bulgaria and Romania the most different. The considerable levels of systemic inertia in some states means little effort has been made to embed MNE, and where efforts have been made, they have been directed to domestic production substitution through mergers with or acquisitions of former state-owned firms. In some cases, the domestic linkages of acquired firms have been maintained; however, in most, many of these linkages have been ousted by the parent multinational's global network of affiliates and partners.

The tendency in most cases, therefore, is a focus on FDI flows with little after-care or efforts to embed, which are necessary for successful FDI-assisted development. While some countries, such as Hungary, the Czech Republic, Slovakia and Romania, have tried to consider FDI and industrial policy in tandem, other countries have made only loose connections and regard MNE activity and industrial restructuring and growth as separate policies. Few countries, even among the core EU economies, see any necessity for a three-way link involving MNE, industry and innovation policies. In our view, the three are inextricably linked.

It is important to emphasize the difference between the advanced and the new NMS, although it seems that neither group of countries is likely to attract significant supply-side R&D foreign investment. With the exception of the Czech Republic and Hungary (OECD, 2008), few NMS have sufficiently well developed science and technology infrastructures to offer absolute advantage in basic research which might attract a MNE to locate a stand-alone, specialized affiliate R&D facility. Indeed, there are only a few locations within the EU as a whole that have adequate science and technology infrastructures. *It is therefore most practical here to recommend that NMS focus on attracting and fostering demand-driven R&D activities by MNEs.* These recommendations, therefore, do not differ greatly from those applicable to the embedding of FDI in general. As with all MNE-embedding policies, the focus must at all times be centred on the deepening of existing MNE value-adding activities, and the promotion of sequential investments that pull the MNE activities such that they become deeply integrated with the MNE global structure and simultaneously deeply embedded within the domestic innovation system. In other words, the goal remains to increase the strategic importance of MNE domestic affiliate for the MNE headquarters, such that sequential investment is increasingly knowledge-intensive. Figure 6.6 shows that the goal is to transform subsidiaries from single-activity units and miniature replicas (quadrants A and B) towards quadrants C and D.

6.5.1 Reducing the Emphasis on Cost Advantages

There is a tendency for many countries to measure their potential attractiveness to MNEs based on their basic infrastructure and relatively low-cost labour. But these kinds of location advantages are generic, in the sense that they are widely available. Furthermore, MNEs do not tend to locate their innovative activities based on cheap factor inputs; if they do, the activity tends to be of the sort that is 'footloose', such as clinical trials for pharmaceuticals (for example Filippov and Kalotay, 2009). In addition, the last two decades of increasing liberalization, falling transportation and communication costs, and investment in knowledge-based activities in East Asia (by both domestic firms and MNEs) has meant that basic infrastructure and low wages are not a magnet for investment. It is axiomatic that as industrial development takes place, the comparative advantage of these countries needs to shift from low value-adding activities to higher value-adding activities, which necessarily are science-based, but the infrastructure (which forms an important quasi-public good) necessary to achieve this is not always available.

6.5.2 Increased Emphasis on Specialized, Location-specific Assets

It is only in those sectors where 'specialized' location advantages associated with higher value-added exist that host countries can benefit significantly from MNE activity in the long run. This requires a considerable amount of government interaction and investment in tangible and intangible infrastructures. As countries reach a threshold technological capabilities level, governments need to provide more active support through macro-organizational policies. This implies developing and fostering specific industries and technological trajectories, such that the location advantages they offer are less 'generic' and more specific, highly immobile and encourage mobile investments to be locked into these assets. Many NMS have the bases for creating such science-based location advantages. For instance, Poland has strengths in certain natural and life sciences, while Hungary has strengths in the electro-mechanical sectors. Of course, adapting to such challenges is not costless from four points of view. First, countries need considerable resources to invest in such vertical industrial policy actions. Many industrial policies to foster new sectors have failed because investment is often limited to building up only one part of the innovation system. For instance, Norway's biotechnology initiative to encourage domestic and foreign firms to undertake R&D (initially resulting in more than 50 new biotech companies within two years in the Oslo area alone) did not maintain its initial momentum since there was lack of investment in PhD programmes in universities in the natural sciences. Second, introducing targeted programmes requires considerable political will and discipline, not just because picking one sector or industry requires others to be given less priority, but also because other industries will necessarily need to be 'wound down'. Third, fostering new sectors requires major institutional change. Such radical systemic change requires resources and an effective period of transition given the inertia associated with informal institutions. Fourth, developing a new sector needs to be undertaken in a 10–15-year time frame.

6.5.3   Creating Clusters around MNEs

One of the challenges in creating MNE embeddedness is matching the industrial structure and comparative advantages of a region with types of FDI. As highlighted above, the benefits of FDI are maximized when the investment projects match with potential domestic competitive clusters which MNEs may be able to tap into.

In many locations, including the US and the EU, efforts are made to attract large projects to act as hubs around which clusters can be built. Large incentives and subsidies are provided in the hope that as well as attracting foreign investors to the region, they will promote substantial linkages and growth in the domestic sector. The investment by Toyota in St Petersburg is a case where sequential investment occurred through other Japanese firms following Toyota's lead. However, little or no attention was paid to the knock-on effects in the form of growth opportunities for domestic suppliers. Although several automotive firms are now operating in the St Petersburg–Leningrad Oblast area, either in collaboration or independently, providing considerable opportunities to pool skills and capabilities, these opportunities have not been exploited. Policy-makers in this case focused only on capital flows and employment, not linkages. Chobanova (2009) notes a similarly passive approach in policies for FDI in Bulgaria and Romania, which contrast with the positive efforts in the Czech Republic and Hungary.

If we look farther afield, there is the example of Costa Rica's success in attracting huge investment from the US MNE, Intel, which became the basis for a sizeable industry of both foreign and domestic firms. The attractiveness of Costa Rica was based more on availability of a skilled and capable workforce than incentives (see Mortimore and Vergara, 2006). Intel's decision to invest in Costa Rica in the mid-1990s had a huge impact on the Costa Rican economy, and represented consolidation of the national strategy to diversify out of apparel and natural resources toward electronics. The investment had a ripple effect in the economy in terms of related activities, such as software, and Costa Rica designed and implemented a new development strategy based on attracting FDI to upgrade into more technologically sophisticated activities. It achieved considerable success in electronics, medical devices and logistics through selective interventions using innovative FDI promotion techniques with the aim of improving domestic capabilities to attract more FDI, through an active and targeted FDI policy reflecting national developmental priorities. It identified the MNEs to be targeted and negotiated firm-level packages, and designed and implemented industry policies to deal with problems that arose, especially in relation to weak technology transfer and assimilation, and limited productive linkages. Costa Rica stands out as an example of what can be achieved by a coupling between an appropriate policy framework that reflects the priorities of the national development strategy, and the global expansion strategy of a leading MNE. It demonstrates how national policy goals and corporate strategy objectives can coincide (Mortimore and Vergara, 2006; Mytelka and Barclay, 2006).

However, attracting FDI through subsidies without trying to maximize the linkages to the domestic economy can lead to a net negative sum game. In 1993, the US state of Alabama gave $253 million in subsidies and incentives to Daimler Benz: it was not until recently that the outlay could be said to have been justified. The European NMS cannot afford to exploit incentives and subsidies without some certainty that there will be potential spillovers and linkages, and that these will be converted to actual benefits.

A number of NMS – the Czech Republic and Hungary in particular – attempted to encourage MNE embeddedness prior to EU accession through the use of broad policy measures: high tariffs and customs duties; rule of origin; local content, and so on. However, after the accession of these countries, many MNEs, even in low-technology sectors such as food and beverages, relocated (Chobanova, 2009). Without some sort of benefits, MNEs prefer the economies of scale and scope in existing activities in the core EU countries despite the low-cost advantages offered by the NMS (Chobanova, 2009). Such import-substitution type policies, therefore, were only a short-term (and short-sighted) strategy, as EU and WTO law requires MNEs to receive national treatment.

6.5.4   Helping MNEs Create Linkages

Policy-makers and consultants are keen to suggest that attracting MNE affiliates associated with global production chains is an important goal. The suggestion is fallacious, since such affiliates are rarely embedded in the local economy, and are deeply embedded in the MNE. It is generally the case that the most embedded affiliates tend to be those with the highest levels of autonomy within the MNE structure, able to make decisions about the nature and extent of their domestic linkages (Castellani and Zanfei, 2006). Affiliates that are responsive to and dependent upon MNE headquarters to make decisions are less embedded, partly because the headquarters have incomplete information on local options and rarely have a vested interest in developing new suppliers. In some exceptional cases, subsidiaries might be well embedded in their local milieux and deeply integrated into the MNE network, but this usually applies only to acquisitions of domestic competitors with historical local embeddedness, which allow the MNEs to inherit ready-made linkages. In other cases, the subsidiary may be very important strategically due to long-term and sustained investment. In both cases, dual integration, while highly desired by governments, is rarely achieved. Chobanova (2009) shows that even in the case of the food and beverages sector, where the skill and knowledge content is quite low, NMS were unable to sustain domestic production by MNEs even after acquisitions of existing firms with proven and fairly well-developed supply chains. Creating local suppliers for MNE networks takes time, and the efficiency of these suppliers cannot be measured only in terms of cost: it must be reliable and flexible relative to international standards. The Czech Republic and Hungary invested heavily in creating linkages between MNEs and domestic firms in the food industry, and upgrading their innovation systems. However, competition from other EU locations did not allow their activities to survive after accession (Chobanova, 2009).

As MNEs increasingly seek to rationalize their activities (as is the case with industries that operate global production networks), decisions about local linkages are not always made at subsidiary level; they are often made at headquarters level, based on comparing the various options available to the MNE globally. Thus, governments need to create incentives for the

MNE to consider local partners, and should not expect this to happen 'naturally'. Since EU member states cannot discriminate by nationality of ownership, in the circumstances where domestic firms are not present, linkages between foreign affiliates and other foreign firms (but located and engaged in economic activity in the same host location) may represent the sole available mode of industrial upgrading and capability development in the NMS. It is important that existing domestic capability is reliable and of high quality, which is often not the case in transition countries. This may be due to poor management, and agencies (for example Ireland) can offer a comprehensive range of services (under strict guidelines and usually within a cost-sharing approach) to assist clients to develop their business strategies, enhance their skills and reduce costs.[3] Enterprise Ireland provides funding for feasibility studies, market research, mentor network service, trade fair participation and training support.

MNEs seek well-established existing location advantages, and initial scale of entry tends to be small in both size and scope. They encompass competence levels that match the innovation system's existing capacity, which in the case of NMS is often lower than the core EU economies. MNEs tend to restrict R&D activities to a few locations, which means that the advantages of a location must be high to attract a new research facility and must offset the costs of exiting from another location. New locations have entry costs related to connecting with local institutions and becoming 'club members' in the innovation system. Initiatives such as the EU's Twinning Programme introduce new entrants to potential partners. For instance, CzechInvest maintains a database of competences and potential suppliers, and introduces potential suppliers to foreign investors as part of its attraction and aftercare service. The Czech Supplier Development Programme run by CzechInvest has been in place since 1999. Its objective is to strengthen contacts between domestic suppliers and MNE manufacturers already operating or planning to invest in the Czech Republic in order to increase the competitiveness of Czech suppliers and help multinational manufacturers identify new partners from among the ranks of Czech suppliers.[4]

There is an important paradox regarding the implementation of industry and innovation policy. MNEs establish and expand affiliates based on existing opportunities to link to the innovation system. However, many countries rely on MNEs to reinvigorate their innovation systems and boost growth through greater linkages. MNE activity will not provide growth opportunities unless there is an existing domestic industrial sector that has the necessary technological capacity to profit from the externalities from MNE activity. In some of the NMS, there are very few viable domestic firms, which limits the possibility for growth.

6.5.5 Improve Opportunities for Start-up and Small and Medium Sized Firms

The policies in many EU member states, including EU core members, focus on large firms only. More than 70 per cent of R&D in the Netherlands, Finland, the UK, Italy, Sweden, Germany and France is carried out by large firms (OECD, 2008). Similarly, these countries tend to target the larger MNE. However, the advantage gained from the presence of a small MNE is that they have smaller intra-MNE networks and are more likely to embed locally.

In the newer industries especially, there is no dominant large firm, and the emphasis must be on establishing start-ups. Singapore's Local Industry Upgrading Programme runs a Startup Enterprise Development Scheme (SEEDS). Start-ups can apply for early-stage SEEDS equity financing. Every dollar raised by a start-up from a third-party investor is matched by SEEDS, up to a maximum.[5] In high-tech (and thus more risky) sectors, Singapore's Enterprise Development Board provides risk sharing in technology-based ventures with investors, via its Enterprise Investment Incentive (Technopreneur) Scheme.[6] This scheme gives qualifying firms insurance against loss of their investment, and extends to foreign firms, although only those incorporated in Singapore. Singapore encourages inventors to patent and commercialize their inventions. The Technopreneur scheme covers some of the patent application filing costs.

6.5.6 Improving Human Resources Capabilities in Line with Demand

Human resource capabilities are important at two levels. First, there are human resource capabilities that are generic to the innovation system, which traditionally have been university graduates. However, modern innovation is demanding a broader range of technical and craft skills. Tertiary education needs to focus on all these levels and develop programmes to respond to industry demand. In Singapore, for example, the Ministry of Trade and Industry, the Economic Development Board and the Council for Professional and Technical Education, work together to monitor future skills requirements, with inputs from foreign and local investors and education and training institutions. This information is matched to national policy objectives and used to focus university, polytechnic, school and technical education (UNCTAD, 2005). Such efforts also require different skills from teachers, trainers and university lecturers, and expertise in basic infrastructure projects such as electrical power generation, construction, and so on. Education is needed in business infrastructure and investment to develop skills such as accounting and actuarial sciences.

Second, there are specific skills required by specialist domestic and foreign firms. Universities and polytechnics should be encouraged to work with MNEs to provide different types of training: on-the-job training for employees to develop skills in particular areas, and specialized training for potential employees. The former should be in collaboration with tertiary institutions. Also, some FDI subsidies require that foreign investors provide some specialized training. Training provision can be in the form of a subsidy per employee (for example 35 per cent of costs, up to a maximum limit of $500, for up to 100 employees per year for specialist training, and 20 per cent of costs for general training). Specialist training refers to skills that are not easily transferable to other companies. They increase MNE productivity and the overall quality of the human capital in the country.

MNEs sometimes invest in developing specialist training programmes in higher education institutions to promote the proper training of potential employees, but these are largely individual firm initiatives, and this is a model that is only viable for large MNEs with special needs, and which can afford this investment (Chobanova, 2009). In many countries governments require firms to provide internships for technical positions, in collaboration with the local technical training schools. This raises the quality of the education institutions and enables firms to identify potential employees. However, the operation of such initiatives is dependent on the goodwill of the foreign affiliates. In the US, South Carolina's Center for Accelerated Technology Training (CATT) programme provides a formal link between employers and technical institutions. CATT is subsidized by the State of South Carolina and helps investors to find appropriately qualified workers while providing some level of specific skills-training for blue-collar workers. It focuses on the training needs of new and existing business and industry in South Carolina. CATT's services are provided through state tax dollars at minimal or no cost to the qualifying client. It specializes in the provision of custom-designed, short- and long-term training for industries seeking to locate to or expand in South Carolina. It provides recruiting, assessment, training, development and management and implementation services to clients that are creating new jobs at competitive wages and with benefits. Training may be delivered through pre-employment or on-the-job activities, depending on the needs of individual customers.[7]

6.5.7 Building Research Capacities in the Public Sector

There are two aspects to building research capacity in the public sector. First, investment in supply-side R&D generation, which includes long-term research projects in specific areas, for example, national laboratories, academies of sciences, and so on. These generate outputs, such as academic publications and patents, which act as important sources of knowledge inputs for the establishment of research establishments by MNEs and domestic firms. Public research institutes also provide technical services for testing and consultancy services for firms as part of the metrology, standards, testing and quality control infrastructure.

Second, there are demand-driven public institutes, which are active in particular sectors, whose primary purpose is to develop specific innovations to meet the needs of particular sectors or groups of firms, and are a quasi-public good. The institute sector in Norway, for example, is comprised of approximately 15 technological research and 30 social research institutes, and reflects the various stages of Norwegian industrial policy since World War II. They can be classified into four main groups. The 'collective' industry-specific research institutes which are based around particular sectoral interests, for example, the pulp and paper industry sponsors the Pulp and Paper Institute. The 'modernization' institutes, which were established in the 1950s as part of the policy strategy to upgrade and develop particular industries were deemed as essential to creating a modern industrial sector. There are regional institutes in Norway, which are linked to local university-level colleges, and support and develop local industry in the Norwegian regions, linking them to the regional tertiary level colleges. And there is a group of institutes that evolved in response to new industries (in particular petroleum and electronics), which have merged into the single organization, SINTEF, which is the largest independent research organization in Scandinavia and employs nearly 2000 people. It undertakes some 60 per cent of R&D outsourced by Norwegian industry. It is organized across eight research areas and controls four stock research companies. The primary reason for the very centralized (and concentrated) nature of the institute sector in Norway is to create economies of scale and scope in research. The Norwegian example is particularly relevant, because most NMS firms are constrained in terms of the resources they can invest in risky R&D, and are unable to afford the high fixed costs of maintaining the R&D infrastructure for occasional projects. If governments establish demand-driven facilities, their services can be shared by large numbers of clients that would not be able to invest in their own R&D facilities.

6.5.8   Policy Tools to Promote R&D

Some countries outside the EU use IPR policies to promote R&D by MNEs, and in China competition policy has been effective in providing large oligopolistic markets for MNEs on condition that knowledge-intensive and R&D activities are undertaken locally, either independently, or in conjunction with domestic firms (Liang, 2007). However, these conditions would contravene EU competition policy and WTO rules since they effectively impose a performance requirement. EU member states are bound by EU regulation on competition and IPR policies that cannot be modified at national level and, therefore, are not available to NMS and are not discussed here. Other options, such as R&D tax credits, lead to greater concentration on innovation activities by firms already active in R&D, although they do not necessarily promote R&D in firms that are not R&D active. Also, if positive and tangible results are to be achieved, the costs of tax credits are often prohibitive. A study by Harris et al. (2008) emphasizes that in disadvantaged regions of the EU, the level of tax credits that would be required would negate the benefits. There would also be strong competition among countries for their receipt. Thus, such broad policy tools are of limited value in the NMS.

6.6   CONCLUSIONS

This chapter has investigated the options available to the governments in NMS to encourage MNEs to invest in R&D. We make the distinction between two types of MNE R&D. Demand-driven R&D is activity undertaken to adapt existing products and services to local needs. Supply-side R&D is innovation conducted in independent R&D facilities which are knowledge-intensive; this type of R&D implies greater dependence on domestic knowledge sources and infrastructure. These two types of R&D require somewhat different approaches and policy options. We focused in this chapter on the MNE and the potential for linkages beyond the traditional spillovers. MNEs engage in a variety of informal and non-equity agreements to engage in knowledge exchange, and taking this into account adds more depth to the analysis. We considered scope and competence at MNE subsidiary level, two aspects that highlight that the tendency to focus on FDI flows is flawed, since knowledge exchange and innovation are establishment-level phenomena. An MNE policy is required to link FDI and industrial policy. We argue that NMS should focus on attracting and fostering demand-driven MNE R&D activities. We recommend also that governments should reduce the emphasis on costs and increase the emphasis on specialized, location-bound knowledge assets, and set up programmes that foster demand-oriented upgrading of public R&D and human capital.

NOTES

1. Malta and Cyprus are not included in the analysis in this chapter.
2. A recent study (Thelwall and Zuccala, 2008) shows that EU cooperation amongst universities still displays considerable divergence. The core EU countries continue to dominate inter-university EU collaboration, particularly the UK and Germany. The new EU members are not well integrated into the wider EU network, although some engage in regional collaborations.
3. See Enterprise Ireland website, http://www.enterprise-ireland.com/Grow/Competitiveness/Build_new_skills.htm. In order to be eligible to apply for these services firms must be a manufacturing or internationally traded services small or medium-sized firm, employing between 10 and 249 people.
4. Czechinvest maintains a database of over 2000 Czech manufacturers/potential suppliers; see http://www.czechinvest.org/en/czech-suppliers.
5. http://www.edb.gov.sg/edb/sg/en_uk/index/our_services/startups/financing/startup_enterprise.html.
6. http://www.edb.gov.sg/edb/sg/en_uk/index/our_services/startups/financing/enterprise_investment.html.
7. Source: http://www.cattsc.com/.

REFERENCES

Applied Research and Communications Fund (ARCF) (2006), 'Innovation.bg. Measuring the Innovation Potential of the Bulgarian Economy', ARCF, Sofia, available at http://www.arcfund.net/fileSrc.php?id=1732, accessed 13 August 2010.

Arundel, A. and A. Geuna (2004), 'Proximity and the use of public science by innovative European firms', *Economics of Innovation and New Technology*, **13**(6), 559–80.

Asiedu, E. (2006), 'Foreign Direct Investment in Africa: the role of natural resources, market size, government policy, institutions and political instability', *World Economy*, **29**(1), 63–77.

Benito, G., B. Grogaard and R. Narula (2003), 'Environmental influences on MNE subsidiary roles: economic integration and the Nordic countries', *Journal of International Business Studies*, **34**, 443–56.

Bruszt, L. and G. McDermott (2008), 'Transnational integration regimes as development programmes', in L. Bruszt and R. Holzhacker (eds), *The Transnationalization of Economies, States, and Civil Societies*, New York: Springer, pp. 23–61.

Castellani, D. and A. Zanfei (2006), *Multinational Firms, Innovation and Productivity*, Cheltenham, UK and Northampton, MA, USA: Edward Elgar.

Chobanova, Y. (2009), *Strategies of Multinationals in Central and Eastern Europe*, Basingstoke: Palgrave Macmillan.

Damijan, J., M. Knell, B. Majcen and M. Rojec (2003), 'The role of FDI, R&D accumulation and trade in transferring technology to transition countries: evidence from firm panel data for eight transition countries', *Economic Systems*, **27**(2), 189–204.

Dunning, J.H. and R. Narula (2004), *Multinational and Industrial Competitiveness: A New Agenda*, Cheltenham, UK and Northampton, MA, USA: Edward Elgar.

Edquist, C. and B. Johnson (1997), 'Institutions and organizations in systems of innovation', in C. Edquist (ed.), *Systems of Innovation: Technologies, Institutions and Organizations,* London and Washington: Pinter/Cassell Academic, pp. 41–60.

Filippov, S. and K. Kalotay (2009), 'New Europe's promise for life sciences?', in W. Dolfsma, G. Duysters and I. Costa (eds), *Multinationals and Emerging Economies: The Quest for Innovation and Sustainability*, Cheltenham, UK and Northampton, MA, USA: Edward Elgar, pp. 41–57.

Fontana, R., A. Geuna and M. Matt (2006), 'Factors affecting university–industry R&D collaboration: the importance of screening and signalling', *Research Policy*, **35**, 309–23.

Garmel, K., L. Maliar and S. Maliar (2008), 'EU Eastern enlargement and foreign investment: implications from a neoclassical growth model', *Journal of Comparative Economics*, **36**, 307–25.

Harris, R., Q. Li and M. Trainor (2008), 'Is a higher rate of R&D tax credit a panacea for low levels of R&D in disadvantaged regions?', *Research Policy*, **38**, 192–205.

Javorcik, B. and M. Spatareanu (2008), 'To share or not to share: does local participation matter for spillovers from foreign direct investment?', *Journal of Development Economics*, **85**, 194–217.

Liang, G. (2007), 'New competition: foreign direct investment and industrial development in China', *ERIM Ph.D. Series Research in Management 47*, Rotterdam: Erasmus School of Economics.

Majcen, B., S. Radosevic and M. Rojec (2009), 'Nature and determinants of productivity growth of foreign subsidiaries in Central and East European countries', *Economic Systems*, **33**, 168–84.

Mortimore, M. and S. Vergara (2006), 'Targeting winners: can FDI policy help developing countries industrialize?', in S. Lall and R. Narula (eds), *Understanding FDI-Assisted Economic Development*, London: Routledge, pp. 53–84.

Mytelka, L. and L. Barclay (2006), 'Using foreign investment strategically for innovation', in S. Lall and R. Narula (eds), *Understanding FDI-Assisted Economic Development*, London: Routledge, pp. 85–114.

Narula, R. (2002), 'Innovation systems and "inertia" in R&D location: Norwegian firms and the role of systemic lock-in', *Research Policy*, **31**, 795–816.

Narula, R. (2003), *Globalization and Technology*, Cambridge: Polity Press.

Narula, R. and C. Bellak (2009), 'EU enlargement and consequences for FDI assisted industrial development', *Transnational Corporations*, **18**(2), 69–89.

Narula, R. and I. Jormanainen (2008), 'When a good science base is not enough to create competitive industries: lock-in and inertia in Russian systems of innovation', *MERIT-UNU Working Papers* 2008-059, UNU-MERIT, Maastricht.

Newman, K. (2000), 'Organizational transformation during institutional upheaval', *Academy of Management Review*, **25**, 602–19.

OECD (2008), *Science and Technology Outlook 2008*, Paris: OECD.

Radosevic, S. (1999), 'Transformation of science and technology systems into systems of innovation in Central and Eastern Europe: the emerging patterns and determinants', *Structural Change and Economic Dynamics*, **10**, 277–320.

Radosevic, S. (2003), 'Patterns of preservation, restructuring and survival: science and technology policy in Russia in post-Soviet era', *Research Policy*, **32**, 1105–24.

Radosevic, S. (2006), 'Central and Eastern Europe between domestic and foreign led modernization', mimeo.

Rodrik, D. (1999), 'The new global economy and developing countries: making openness work', *Policy Essay no. 24*, Washington, DC: Johns Hopkins University Press for the Overseas Development Council.

Rodrik, D., A. Subramanian and F. Trebbi (2004), 'Institutions rule: the primacy of institutions over geography and integration in economic development', *Journal of Economic Growth*, **9**, 131–65.

Sinani, E. and K.E. Meyer (2004), 'Spillovers from technology transfer: the case of Estonia', *Journal of Comparative Economics*, **32**(3), 445–66.

Thelwall, M. and A. Zuccala (2008), 'A university-centred European Union link analysis', *Scientometrics*, **75**(3), 407–20.

UNCTAD (2005), *World Investment Report 2005*, New York and Geneva: United Nations.

7. Innovation in EU CEE: the role of demand-based policy

Jakob Edler*

7.1 INTRODUCTION

The 2007 Innovation Report for Bulgaria, which is based on a survey of highly innovative Bulgarian companies, draws conclusions about how demand is a major bottleneck:

> Together with their strong commitment to innovation, the companies from this study mention a list of problems which explain why the number of innovative enterprises in Bulgaria is so low. In general, there is limited local demand for innovation products. Bulgarians do not easily accept new technologies and even those companies that recognize the need for innovation and modernisation of production and would like to invest in R&D have problems realizing their goals. (Ministry of Economy and Energy, 2007: 94)

This is followed a couple of lines later with a request for supply-side measures, but not support for marketing: 'In summary, the main barriers before innovative companies are the insufficient human resources, the lack of financial mechanisms for high-risk projects, the lack of resources for active marketing and tax incentives as well as the need to create a favourable environment for start-up companies' (Ministry of Economy and Energy, 2007: 94–95).

These quotes are illustrative of the poor demand for innovation in many parts of the European Union (EU) Central and East European (CEE) economies, and show that although there is some recognition of a bottleneck, the solution is most often sought in supply-side measures.

The economies and the innovation systems of the CEE countries are still in a state of transition and catch-up. In the transformation of their innovation systems, one of the major characteristics of those European countries with a socialist past has collapsed, that is, the old production structures and the systems for the application and exploitation of knowledge. In their efforts to re-build their innovation systems, EU CEE countries are applying different kinds of strategies.

Despite many differences, it appears that all CEE countries are focusing on innovation supply and paying little attention to demand. This is in line with the policy rationales and academic discussions that have prevailed across the OECD countries since the late 1980s. The question of how demand conditions affect the productivity and innovation dynamics in CEE has not been discussed systematically, let alone considered in policy strategies. This is despite the fact that fostering demand for innovation can produce three interdependent effects: (1) leading edge demand can incentivize suppliers to produce innovation; (2) the absorption and application of innovation in industry can severely increase business productivity; and (3) procurement and application of innovations in public services can contribute to the achievement of societal goals, potentially improving the performance and responsiveness of the public sector and the welfare of society.

This chapter discusses demand, demand conditions and demand-based policies for innovation in the context of the EU CEE countries. Far from regarding demand policies as a simple silver bullet, this chapter argues that the EU CEE countries' catch-up could potentially be accelerated through complementary demand orientation. At the same time, this could be linked to more satisfactory fulfilment of society and industry needs. The main argument proposed in this chapter is that institutional adaptations, and a policy mix that tackles the bottlenecks to demand for innovation and supports the articulation of demand, can link the modernization of the economy and public services to incentives for innovation activity in the EU CEE country economies and contribute to a tailored catching-up process.

The chapter first presents a short conceptualization of demand-based innovation policy and introduces a typology of demand measures. Section 7.3 discusses demand conditions in the EU CEE countries, linking the discussion to concrete policy challenges. It summarizes some illustrations for the (still very poor) trends towards the demand side in policy design and implementation in the EU CEE countries. Section 7.4 summarizes the main arguments and policy challenges and proposes a set of policy recommendations for achieving a broader, more balanced mix of policies better suited to tackling the specific situations in the EU CEE countries.

7.2 THE CONCEPT OF DEMAND-BASED INNOVATION POLICY[1]

7.2.1 Context and Definition of Demand-based Innovation Policy

Although the various versions of the national system of innovation (NSI) approach (for example Lundvall, 1992; Edquist, 1997) include demand as a key innovation category (users of the new knowledge and customers for the innovations), the demand side has long been neglected in innovation systems analyses and in concepts and innovation policy practice in the OECD countries. The triangle of policy-makers, policy and innovation analysts, and the business community has paid little attention to stimulating demand through innovation policy, where the focus has been supply-side strategies and activities in which research and innovation policy are differentiated.

In this chapter, *demand-based innovation policies (DBIP)* are defined as the set of public measures designed to increase demand for innovations, improve conditions for the uptake of innovations and/or improve the articulation of demand in order to spur innovation and the diffusion of innovations. This broad definition of DBIP signals is aimed at spurring innovation *as well as* the uptake and diffusion of innovation. The emphasis on the diffusion of innovation is a sign that the concept of innovation is not confined to 'new to the world', but also encompasses new to a firm or a particular geographic space. This is of major importance for countries that are not at the forefront for producing new technology and innovations. Further, the definition also contains the articulation of demand, which highlights the importance of defining and expressing societal demands and capturing and translating them into articulated market demands.[2]

The underlying assumption in this chapter is that demand policies should complement, rather than substitute for supply-side measures. Thus we do not need to engage in the somewhat academic debate on whether it is more important to support supply or demand in order to spur innovation and subsequent productivity and competitiveness, or to discuss the various models to understand how demand can influence innovation. Suffice it to say that there is an abundance of literature stressing the importance of demand and challenging the demand conditions and diffusion patterns that incentivize innovation, reduce uncertainty for innovators (for example Schmookler, 1962; Mowery and Rosenberg, 1979; Fontana and Guerzoni, 2007), provide input to the innovation process (for example von Hippel, 1986) and contribute to the productivity and competitiveness of firms and markets (more generally Porter, 1990; Edquist et al., 2000a; McMeekin et al., 2002; Bhide, 2006; Anderson, 2007).[3] Hollanders and Arundel (2007) show empirically that the set of variables that characterize demand conditions are correlated with the innovation performance of countries, two moderately (public procurement, demanding standards), and three strongly (trust, corruption, technology absorption). Hollanders and Arundel's analysis thus emphasizes the meaning of demand conditions for innovation.[4]

Against this background, we can describe the rationale for demand-based *policy* as resting on four pillars (for a more detailed account see Edler, 2010).

1. Innovation policy: overcoming system failures. As in the case of supply orientation, there is a range of system failures that are manifest in a set of bottlenecks that justify policy interference:

    - information and adoption problems in innovation markets (Cohen and Levinthal, 1990);
    - lack of mutual understanding between producer and user (von Hippel, 1986), resulting in human and social needs not being automatically translated into clear market demands (Mowery and Rosenberg, 1979);
    - high entry costs for innovations detrimental to initial uptake, blocking future scale and network effects, and producing lock-in and path dependency. This can impede socially desirable reorientation to new technological trajectories: the stronger the network effects from the innovation, the stronger the case for policy intervention.

2. Societal goals and policy needs. The second justification for demand-based innovation policies is serving societal needs, since public sector innovation activity leads to more effective and more efficient public services.

3. Industrial/economic policy: diffusion and absorption. Bhide (2006) argues that downstream activities and the *diffusion* of technologies potentially results in greater economic effects through productivity gains (and multiplication and network effects) than the original *production* of innovation. Bhide claims that national economic advantage rests on the ability quickly to absorb and make use of new technologies ('venturesome consumption'), rather than being the locus of innovation and research and development (R&D) (Bhide, 2006: 18). These claims are supported by data from Eaton and Kortum (1995), who find that growth in the UK, West Germany and France is based much more on the results of foreign rather than domestic R&D.

4. Industrial/economic policy: pushing local innovation production and creating lead market potential. Demand-oriented policies are re-emerging as industrial policy tools. The extreme argument is that creating lead markets with upfront demand and related domestic production can lead to competitive advantage in domestic and foreign markets. Supply-side measures alone cannot achieve this; producers seek to be close to lead users and markets with leading edge demand

(Jacob and Jänicke, 2003; Beise et al., 2003; Meyer-Krahmer, 2004; Edler and Georghiou, 2007).

7.2.2 The Policy Instruments

There is a wide set of demand-based innovation policy instruments (for more detail see Edler, 2010, 2007a, b, c). The role of the state in influencing innovation through demand is that of a purchaser, and public procurement is receiving increased attention (Wilkinson et al., 2005; Georghiou, 2007; Edler et al., 2005, Edler and Georghiou, 2007). The principal objective is to direct public purchasing towards innovations. In the 1970s and 1980s, several empirical studies concluded that over extended time periods, public procurement triggered greater innovation impulses, and in more areas, than did R&D subsidies (Rothwell and Zegveld, 1981). Dalpé (1994) and Dalpé et al. (1992) show that, especially in research-intensive fields, public administrations are often more demanding than industry or other consumers (Dalpé, 1994). This leads to the conclusion that procurement policy is far more efficient at stimulating innovation than any of a wide range of frequently used R&D subsidies (Geroski, 1990: 183), since administrations are more often 'lead users' for new innovations than private consumers (Dalpé et al., 1992: 258ff). Public procurement can also be linked to and deliberately trigger *private* demand (catalytic procurement) especially if the state bundles demand with companies or private consumers (cooperative demand). The latter two variants have been successful, for example, in programmes to transform energy markets towards more eco-efficiency (particularly well implemented and documented for Sweden: Neji, 1998; Suvilehto and Överholm, 1998).

A second major pillar of DBIP is *support for private demand*, most directly through demand subsidies in the form of direct payments, tax credits or exemptions, or some other financial advantage attached to the use of an innovation.

Further state activity may systematically *improve demand competence* as one key driver of the diffusion of innovation (Gatignon and Robertson, 1985). Following Porter (1990) and several other empirical studies (for example see von Hippel, 1986), the more capable the users, the more likely they will demand and adopt innovations. Policy may improve awareness (marketing support, state as lead user), skills (training, education, demonstration) and transparency (for example labelling, demonstration projects) and shape markets and applications through all kinds of regulations and support for standard setting (see also Blind, 2007).

Even more fundamental, the state can organize and support the *articulation of demand*, not only in reaction to existing innovations (risk discourse), but also in terms of trying better to understand societal preferences and their development and how they link to technological trajectories. Societal preferences are often not translated sufficiently well into signals to the market place – and needs are not heard, or technologies that do not reflect the risks they encapsulate might be pushed into the market.[5] Both call for public action to support articulation of demand (Smits, 2002).

Finally, there is a *range of combinations of measures*. Demand-side measures may be deliberately linked to other demand-side measures and/ or to supply-side instruments. We follow the argument in David et al. (2008: 16) that 'public policy supporting innovation has proven to be especially effective where funding for R&D was combined with complementary policies supporting the adoption of innovation'.

This variety and complexity of instruments points to the immense governance and coordination challenges involved in DBIP. These challenges are a major reason why demand for innovation has not been at the forefront of innovation policies in EU CEE countries. We return to this later in the chapter. It is important also to understand that the role of the state differs among the different instruments: the state may be (only) a financier, or also a moderator, a facilitator and a trainer.

7.3   THE ROLE OF DEMAND CONDITIONS AND RELATED INNOVATION POLICY IN EU CEE COUNTRIES

7.3.1   Innovation Capabilities: A Challenging Starting Point

In terms of innovation supply, the European Innovation Scoreboard (EIS)[6] (which is composed mainly of supply-side indicators) shows that EU CEE countries by and large are lagging behind. However, there are significant differences among EU CEE countries, with some already 'moderate innovators', with only a relatively small gap with the more advanced economies, while others are having to catch up from a low level.

This situation constitutes an important framework condition: policy attempting to articulate sophisticated demand and to spur demand for innovation unfolds within a challenging supply-side context. By and large, this means that supporting demand for leading edge innovations might trigger demand for products and services produced outside the country or by foreign actors. This often raises concerns related to the use of DBIP instruments and must be acknowledged in the design and selection of demand-based policy measures.

Radosevic (2004a) provides an indicator-based comparative analysis of the EU CEE countries based on data for the late 1990s–2000. This analysis goes beyond the limitations of the EIS and integrates a range of variables into the four sub-indices: absorptive capacity (mainly level and breadth of education); R&D supply; diffusion (in Radosevic's model mainly information and communication technology (ICT) diffusion and uptake indicators plus ISO9000 certification); and demand (market pull) variables, to provide an overall National Innovation Capabilities (NIC) indicator.

Not surprisingly, this NIC index shows that the EU CEE countries are behind the EU15 average and that countries are heterogeneous (but less so than the then EU15). Across all indicators, the EU CEE countries are not one distinct group, and include the Southern EU countries (Radosevic, 2004a: 653). In terms of absorptive capacity in particular and the diffusion variables, the EU CEE countries are very diverse. Radosevic shows that 'the *capacity to generate demand* for innovation is the *weakest aspect* of the national innovation capacity of the EU CEE countries' (Radosevic, 2004a: 655 – emphasis added). This is an important finding: apparently the conditions for demand to act as a stimulus to innovation are generally poor, with Hungary, Estonia and to some extent Czech Republic being slightly better placed.

This analysis is pertinent to the topic of this chapter in providing some basic context for demand policies and showing that EU CEE countries are not a unique bloc. However, it shows also the limitations of traditional innovation capability indicators when the focus is on demand and the potential of demand-based innovation policy. The variables in Radosevic's study of demand conditions are development of the financial system (stock market capitalization in percentage of gross domestic product – GDP, domestic credit provided by the banking sector), level of competition (percentage of foreign direct investment – FDI, share of trade in GDP, index of patent rights) and macroeconomic stability (unemployment, consumer price index), where some EU CEE countries perform well.[7] The underlying assumptions are that a well developed financial market, high levels of competition and good macroeconomic stability contribute to greater demand for innovation. The index of absorptive capacity in Radosevic's study is based largely on education and employment in the high- and medium-technology sectors, on the basis that better skills and a workforce engaged in the high-technology industries provide a foundation for catch-up by the economy. These assumptions are not contested, but if the interest is in real potential and the prerequisites for policy to actually spur innovation by influencing demand and demand conditions, it is necessary to go further in terms of both absorptive capacity and demand conditions.

A good start has been made within the Innovation Watch initiative in the context of Europa Innova (Bruno et al., 2008), with a new set of variables compiled from various sources. The idea is to understand and characterize countries' cultural capital and consumer behaviour. The scores of CEE countries in the so-called overall cultural capital index based on seven variables[8] show that, on average, the EU CEE countries[9] are below the EU mean. Very simply put, this indicates a lower tendency to undertake innovation activity. The only country above the EU average is Hungary. The analysis also shows the heterogeneity among countries in their attitudes to science, risk and innovation.

Absorptive capacity is an indication of the keenness of firms' decision-makers to invest in new technologies, to build the necessary absorptive capabilities, that is, to be able to adjust their processes and workforce to the requirements imposed by the new technologies. Training and education are the prerequisite for the production and use of innovation. The related innovation management needs to be understood more broadly. Even more important, absorptive capacity in the public sectors of EU CEE countries is an issue: are public administrations willing and able to demand and utilize innovations and act as lead users of innovation? This links directly to the more general demand conditions: what are the conditions that producers of innovation face when they enter an EU CEE country market, and how dynamic and leading-edge is the demand in these countries? Is it sufficient to send signals to producers about potential innovations?

7.3.2   Demand Conditions in EU CEE Countries – and the Policy Challenges: A Closer Look

7.3.2.1   The spectrum of demand conditions

Within the EU, our data and intelligence on conditions for innovation demand are poor. Despite the studies cited (Hollanders and Arundel, 2007), we lack broad and comparable data on buyer sophistication, scale of innovation in public procurement (and even comparable data to define the share of public purchasing in GDP in EU countries), and the factors shaping the patterns of diffusion of new technologies and innovations. Their absence is a clear sign of how much academic and policy discourse is oriented to the supply side.

One source of information to characterize demand conditions across different countries is the World Competitiveness Report produced by the World Economic Forum (WEF).[10] The data in this report are based largely on surveys of business leaders who offer subjective assessments of a series of supply- and demand-side variables. While these are not hard data, they show how business decision-makers *in the actual countries* perceive the situation: these perceptions can be the basis for decision-making.

The WEF data appear to be comprehensive, international and regular and look at demand conditions in more detail; many of the indicators used show how innovation systems demand innovations, and assess the quality of the framework conditions (WEF, 2008, for the methodology see pp. 68–70).[11]

Among the range of variables in the WEF report, we focus on a selection that is important to characterize demand conditions, and on a series of conditions that are important for providing a context to demand-based policies since they characterize the market and the nature of (potential) demand and its satisfaction more generally. Table 7.1 presents the variables.

Figure 7.1 gives a first overall comparison of the mean values for the various key variables for the EU CEE group of countries and the remaining EU17 countries. Best performance is indicated by the highest single value for any EU country for each variable. The main message is that there is an obvious gap. The data also show that the EU CEE countries are not a uniform bloc, but are heterogeneous (Table 7.2). In some variables, individual countries are within the EU average, but others lag far behind.

The next section analyses these differences and highlights the heterogeneity among the group of CEE countries providing the basis for the subsequent discussion.

7.3.2.2 Private demand for innovation

A key variable in the importance of demand for innovation is buyer sophistication, which signals the inclination and ability of buyers to select products based on performance rather than price and, thus, their willingness to purchase innovative products and services and to bear their higher costs at the beginning of the life cycle. The more sophisticated are the buyers, the more likely they will be able to judge performance and to weigh the higher entry price of an innovation and the learning and adaptation costs involved, against increased performance and long-term value for money. It is also a proxy for innovation curiosity in the market. A comparative advantage of leading countries (for example Finland) is based on the high buyer sophistication of the population, which, in Finland, is keen to try new things, and to be at the leading edge in adopting new technologies (see for example Ebersberger, 2007). Buyer sophistication has been determined statistically as one of the main drivers of innovation (Hollanders and Arundel, 2007: 22).

Buyer sophistication in the EU CEE countries is lower than the EU average (Table 7.2). The average on a scale of 1 to 7 for the EU CEE countries is 3.6 compared to 4.86 for the EU17 countries. The WEF data allow for a wider assessment beyond the EU, including 134 world countries. Of

Table 7.1   Variables to describe key market conditions as a basis for demand-based policies

Variable	Description
Demand conditions	
The nature of private demand	
Buyer sophistication	Buyers in your country make purchasing decisions (1 = based solely on the lowest price, 7 = based on a sophisticated analysis of performance attributes)
Firm-level technology absorption	Companies in your country are (1 = not able to absorb new technology, 7 = aggressive in absorbing new technology)
Role of public demand	
Government procurement of advanced technology products	In your country, government procurement decisions result in technological innovation (1 = strongly disagree, 7 = strongly agree)
Favouritism in decisions of government officials	When deciding upon policies and contracts, government officials in your country (1 = usually favour well-connected firms and individuals, 7 = are neutral)
Selected key framework conditions characterizing the market and the nature of demand satisfaction	
Nature of competitive advantage	Competitiveness of your country's companies in international markets is primarily due to (1 = low-cost or local natural resources, 7 = unique products and processes) *Note:* signals the overall assessment of sophistication of markets
Degree of customer orientation	Customer orientation: firms in your country (1 = generally treat their customers badly, 7 = are highly responsive to customers and customer retention) *Note:* signals how suppliers react or would react to an upgraded, more sophisticated demand for innovation
Availability and use of latest technology	In your country, the latest technologies are (1 = not widely available or used, 7 = widely available and used) *Note:* indicates the lack of technology within the country (no matter from which origin) and thus is a (weak) proxy for technological sophistication
FDI and technology transfer	FDI in your country (1 = brings little new technology, 7 = is an important source of new technology) *Note:* shows the degree to which companies from abroad actually fill the gap of technology supply in the countries; in terms of demand-oriented policies technology transfer through FDI is obviously preferable to import, see main text for further explanation
Imports as a percentage of GDP	Imports as a percentage of GDP (hard data, not WEF survey)[a] *Note:* shows how much the countries rely on products/services produced abroad when satisfying their demand

Source: [a]   Based on Economist Intelligence Unit, Country Data Database (May 2008); World Bank (2007); national data (WEF, 2008: 473).

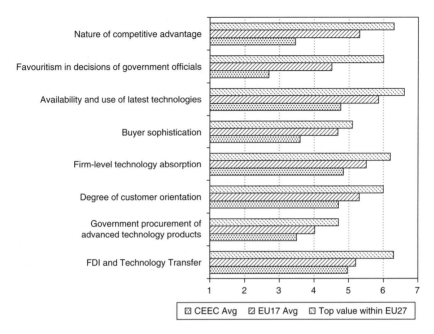

Note: Table 7.1 gives definitions of the variables. Values range from 1 to 7; the higher the value, the more innovation-friendly the condition. This is also true for favouritism: the higher the score, the more neutral is the public contracting. The variables are sorted, so the top variable shows the highest gap of EU CEEC to EU17 mean, the lowest variable the lowest gap.

Source: World Economic Forum (2008), own compilation.

Figure 7.1 *Assessment of context conditions for demand and demand-based policies in EU17 and EU CEE countries*

the countries surveyed, the best ranked in the EU CEE country group is the Czech Republic (41), and the lowest ranked are Bulgaria (87) and Hungary (95). The only countries *outside* the EU CEE countries group ranked *within* the range of the EU CEE countries are Malta (54) and Greece (51). Apart from those two exceptions, the contrast in buyer sophistication in non-EU CEE countries is stark, with 12 of the EU17 countries ranked among the top 22 in the world.

These data show that buyers in EU CEE countries are generally neither inclined nor able to buy leading edge technology and innovation. Thus, the pre-conditions for innovations to be absorbed in the market are challenging. One reason for low levels of buyer sophistication is lack of disposable income among private consumers; however, we need some further

Table 7.2   Selected variables important for demand and demand-based policies: WEF CEE country data (1 = worst score, 7 = best score)[a]

	CZ	EE	HU	LV	LT	PL	SK	SL	RO	BG	CEE COUNTRIES Avg.	EU17 Avg.
Nature of competitive advantage	3.7	3.5	3.5	3.3	3.6	3.6	2.9	4.5	2.9	3	3.5	5.3
Degree of favouritism in decisions of government officials[a]	2.5	3.5	2.4	2.9	2.9	2.5	2.3	3.2	2.4	2.4	2.7	4.5
Availability and use of latest technologies	5.1	5.8	4.7	4.7	5	4.4	5.1	5.1	3.9	3.8	4.8	5.9
Buyer sophistication	4.1	3.8	3.1	3.5	3.7	3.7	3.5	4	3.6	3.3	3.6	4.7
Firm-level technology absorption	5.4	5.5	4.7	4.5	5	4.7	5.4	4.9	4.4	4	4.9	5.5
Degree of customer orientation	4.8	5.3	3.9	5.4	4.7	4.6	4.7	5.1	4.1	4.4	4.7	5.3
Government procurement of advanced technology products	4.3	4	2.9	3.2	3.4	3.7	3.2	3.4	3.5	3.4	3.5	4.0
FDI and technology transfer	5.5	5.3	5.5	4.8	4.7	4.9	6	3.8	5	4.1	5.0	5.2
Import as % of GDP	74.3	81.7	77.7	64.7	67.4	43	86.8	73.2	44.6	85.5	69.89	53.82

Notes:
For a description of the variables see Table 7.1; the ordering follows that in Figure 7.1.
[a]   A higher score means *less* favouritism.

Source: Based on WEF (2008).

investigation in this area. Low levels of sophistication generally require demand policies that target awareness, attitudes and skills. Demand-oriented policies need to influence the innovation culture, make buyers more willing to take risks, increase awareness of innovation, and empower consumers to use them. They need to include large premiums for the consumer (some sort of demand subsidy) to overcome resistance to new products and processes. All of these measures are vital for countries that want to become markets for innovation.

We also analyse the readiness of firms to adopt the latest technology, which is an indication of absorptive capacity for new technologies.[12] The business-to-business market is important for the uptake of innovation from the original suppliers, to provide a return on the supplier's investment in innovation. Adoption of technology is key to the modernization of industry through the rapid diffusion of sophisticated new services, equipment and parts, which simultaneously affect supply and demand. The importance of this variable for the overall innovation performance of countries cannot be overestimated, it is one of the four (in 26) variables that are statistically strongly correlated to the summary innovation index (Hollanders and Arundel, 2007: 35).

The results of the survey show that this variable scores slightly better than the assessment of general buyer sophistication: the mean for the EU CEE countries is 4.8 compared to 5.5 for the remaining EU17. In a global perspective, the EU CEE countries are lagging, the highest ranked country in the group being Estonia (ranked 30th in the survey), ranked 13th among the EU countries, followed by the Slovak Republic (37th in the overall survey) and the Czech Republic (38th in the overall survey). The EU CEE countries whose companies are the least ready to adopt the latest technologies are Romania (94th in the world ranking) and Bulgaria (114th).

This variance again demonstrates that different policies are needed, for instance, for Estonia compared to Bulgaria or Romania, to encourage firms to demand the latest technology. For EU CEE countries, such as Bulgaria and Romania, the barriers to the diffusion of modern technologies are apparently very high, and innovation policy which sets highest standards for the uptake of technology might create deadlock rather than triggering catch-up. The solution is broad awareness and enabling measures, and incentives to overcome short-termism. At the same time, we would expect that in the more advanced countries, such as Estonia, diffusion and adoption policies could be more demanding, targeting actors at higher levels of innovation capabilities and awareness. Nevertheless, for the EU CEE countries as a whole, the data point to the limitations of only supply-side policies, and measures to boost the innovation output of companies that equip and deliver to other businesses will need to be complemented by policies that increase firms' inclination and readiness to modernize by adopting the latest technology.

7.3.2.3 Public procurement for innovation

The above analysis shows that the EU CEE countries, to various degrees, represent 'immature' environments for innovation demand, meaning that the private sector is generally not very willing or able to procure and adopt innovations (Edquist et al., 2000b: 304).

In light of comparatively weak buyer sophistication and lower levels of readiness among companies to demand innovations, can and does public demand fill the gap, and play the lead role in innovation adoption? Although the leverage of public procurement overall is undisputed, we lack solid, internationally comparable data on the relative weight of public demand in the various European countries at local, regional and national levels. The WEF survey asked specifically about whether government procurement decisions result in technological innovation (1 = strongly disagree, 7 = strongly agree). Based on the assessments of business leaders, the first important finding is that public procurement is not generally regarded as driving innovation in either the EU17 or the EU CEE countries: the averages are low (4.0 and 3.5 respectively).

While the CEE countries are lagging, the gap with the EU mean is smaller (see Table 7.2). The perception of public demand for new technology in some leading EU CEE countries is well within the average for the EU countries as a whole, but again there are huge disparities. The EU CEE country that comes out highest in the assessment is the Czech Republic, which is ranked 18th globally, and 6th among the EU countries, with a score above the average for the EU17 group. The second highest ranked EU CEE country is Estonia; the lowest ranked is Hungary (116), which scores the same as Italy and slightly higher than Greece.

According to the WEF business leaders surveyed, the Czech Republic and Estonia have already caught up with some of the leading EU countries. However, most other EU CEE countries have quite a way to go to use regular procurement for innovation and to add innovation demand to their policies. We need to keep in mind that the absolute score for this variable is low for all the EU countries. This points to a set of obstacles that will not encourage public demand for innovation in the EU CEE countries.

The first challenge is related to governance and coordination issues in public procurement, which the EU CEE countries and EU17 generally share although they are more pronounced in the former group. Public purchasing to trigger innovative action in companies or to enable the diffusion of new products or services ideally should combine efficient purchasing (value for money) with better public service or achievement of sectoral policy goals (functional specifications, improved performance). In order for this link to be completed, policy and administrative practice must overcome many organizational, incentive and skills challenges (see Edler and Georghiou, 2007 for more detail). It calls for joint strategy making including the definition of public demand to ratchet up innovation policy and improve well established coordination mechanisms. Further conditions conducive to public purchase of innovations are willingness to take risks, procurers with sound market knowledge, an incentive structure that encourages risk-taking and provides experienced and sophisticated risk management advice, and political decision-makers who uphold the higher entry costs of innovations in the interests of life-cycle costing[13] and improved functional performance.

All these governance conditions are important for all countries, but as all reviews and analyses show (see ERAWatch – EW and PROInno Trendchart reports), most EU CEE countries are still in the stage of transition, institution-building and learning. In these countries (and also in many non-EU CEE countries) we find high levels of administrative fragmentation accompanied by low levels of inter-administrative coordination. The ability to set up sophisticated procurement approaches that meet all the requirements is a long way off for the EU CEE countries. Although this is undisputed at the national level, there are some encouraging examples at the local level, where the much simpler governance structure and higher level of awareness and skills enable more advanced procurement activities (for example in the Baltic region; see below).

The second set of challenges is related to transparency and fairness. Again, they are not unique to the EU CEE countries, but are more pronounced in this group compared to the average EU17 country. The key requirements for public procurement for innovation include a systematic, transparent, open and objective procurement process. To incentivize companies to invest in innovation activities in the context of public procurement they need to feel that the competition is based on clear criteria and that the best value for money ratio is achieved. Corruption, bribery and favouritism in public purchasing decisions are severe hindrances to innovation procurement, especially since market newcomers face high risks from unfair competition. Without open access and entry to the public market, competition will be less innovative. It is not a statistical artefact that the variables 'trust' and 'corruption' comprise two of the four variables (out of a total of 26) that are strongly correlated to innovation performance (Hollanders and Arundel, 2007: 35).

The question posed by the WEF survey was: 'When deciding upon policies and contracts, do government officials in your country usually favour well-connected firms and individuals (=1) or are they neutral (=7)?'. The

EU CEE countries score much lower with an average value of 2.7 compared to 4.5 for the EU17 (Table 7.2). The best ranked among the EU CEE countries is Estonia (44th in the world), ranked 16th among the EU countries, followed by Slovenia (62nd in the world), Lithuania (81st) and Latvia (83rd); the lowest assessed country is the Slovak Republic (116th out of 134 world countries, and the lowest ranked EU country).

This WEF finding is confirmed by the World Corruption Index compiled by Transparency International.[14] On a rating from 0 (highly corrupt) to 10 (not at all corrupt) the EU17 (that is excluding the EU CEE countries) scored an average of 7.44, while the EU CEE countries scored 5.02.[15] Thus, the EU CEE countries have a bad image in terms of fair public procurement practices.

7.3.2.4 Meeting increased demand

In discussing the role of demand for innovation, we need to take account of the interface between supply and demand in terms of both purchasers and suppliers, and early signals from customers. A prerequisite for demand-based innovation, therefore, is that companies should listen to customers, be oriented towards their needs and, with their help, start to anticipate and define future needs. The responses to the question about how companies treat their customers ('treat their customers badly' (1) to 'highly responsive to customers' orientations' (7)), provide a mixed picture. The variation among countries is huge. Lithuania (21st in the world), Estonia (24th) and Slovenia (30th) are among the world top 30 countries, and are ranked 11th, 12th and 13th among the EU countries. The laggards are Romania (100th in the world) and Hungary (114th). Data from the European Community Innovation Survey gives a slightly more encouraging picture. In the EU CEE countries[16] the share of companies reporting users as a source of innovation is as high or higher as for the rest of the EU (the exception being Poland; Bruno et al., 2008: 34). This more open interface between producers and users is a potential asset for demand-based policies.

We are also interested in whether the latest technologies in countries are widely available and used. This is a hybrid variable, which indicates both supply and demand: a high score means the economy is able to provide the latest technology and is willing also to absorb it (see definition in Table 7.2). Again, Estonia is the highest ranked country in the EU CEE countries (21st overall, and 12th among the EU countries). The other EU CEE countries lag far behind, with the Slovak Republic next highest at 46th in the world and Bulgaria ranked lowest (103rd).[17]

Finally, if we assume that local producers do not have the capability to produce the quality and quantity of innovation that is required, we need to enquire about the role of imports and FDI in technology transfer. The idea is that increased demand for innovations locally within the EU CEE countries can take advantage of FDI, bringing in technology or triggering more intensive technology transfer through FDI, which will work to upgrade the system. The short-term, counter argument is that higher demand for innovation might lead to value-added abroad if demand cannot be satisfied through FDI, and innovations are imported.[18] We return to this issue below.

Hard data on the percentage of GDP that can be allocated to imports shows that all EU CEE countries except Poland and Romania (the biggest CEE countries) are above the EU average, and dependency on imports is high. In terms of innovation and, thus, economic policy, policy-makers may want to favour FDI over imports. However, this requires that inward FDI should be technology-intensive and provide for technology transfer (see Chapter 6 in this volume). The assessment of companies in terms of whether FDI is a source of very little (=1) or a great deal of new technology (= 7) shows that there are two distinct groups of EU CEE countries. In four EU CEE countries, foreign firms are contributing considerably to technology transfer, with the Slovak Republic ranked second highest among all EU countries and the Czech Republic, Hungary and Estonia, ranked 5, 6 and 8 among the EU countries. The remaining EU CEE countries are ranked much lower, with Bulgaria and Slovenia lowest among the EU27.[19] Therefore, while in some countries increased demand for innovation could build on strong FDI that is potentially able and willing to deliver new technologies, other countries will struggle and must rely on imports or the building of indigenous capabilities.

As emphasized above, the potential supply bottleneck will be a severe issue for demand-oriented innovation policy *only* if we neglect the benefits of technology and innovation diffusion for societies and economies. Innovation demand and the application of the latest technologies trigger modernization effects at home, in both firms and public administrations (Bhide, 2006). In addition, applying the latest technology, whether bought locally or imported, can produce spillover effects to the local economy through demand for maintenance, service or complementary products, all of which may be provided by local firms, and will motivate local firms to develop new skills.

7.3.3 Policy Practice: Some Observations

7.3.3.1 Lack of systematic demand orientation

Against this background of poor demand for innovation and poor demand conditions more generally, we need to examine the role of policy to spur demand. Based on reports and data from EU Trendchart,[20] EU ERAWATCH,[21] Policy Mix[22] and OMC (Open Method of Coordination) peer review reports for the EU CEE countries (Estonia – Polt et al., 2007; Lithuania – Edler et al., 2007; Bulgaria – Edler et al., 2008; Romania), it is clear that the policy mix in research, technology and innovation in the EU CEE countries does not take systematic account of the meaning of demand conditions for the dynamics of the innovation system. To improve demand for innovation, whether public or private, seems not to be a strategic priority. This is true for all the EU CEE countries, and especially since they need to catch up in many dimensions. Document analysis reveals that 'demand' is still largely understood as company demand for (public) research or demand for skilled labour, but not as demand for innovative products and services.

The modernization programmes within the implementation of the Structural Funds Operational Programmes seem to have missed an opportunity. When screening the types of innovation-related activities to fund, the Operational Programmes do not explicitly support purchase and application of leading edge innovations; rather they support innovation projects in companies very often by subsidizing purchases of assets and machinery.[23] The potential to link the demand for modernization to improved internal innovation capability is not being realized. While, as in the Czech case, the purchase of used machinery is excluded from funding,[24] there is most often *no* premium, no specific incentive to buy innovations, for example, machinery or services, that are new to the market.

Thus, at the strategic level, Radosevic's (2004b) finding is still valid and has been confirmed by the policy reports: the EU CEE countries have focused on high technology, on re-creating their science base and research capacity in the private sector. They are also focusing on re-establishing the link between these two aspects, and the strategy discourse in the EU CEE countries is still very much driven by a science-based, high-tech model that considers diffusion, but usually as a secondary condition.

7.3.3.2 Policy practice at the instrumental level

While in strategy documents the demand side tends to be ignored, at the level of individual instruments innovation policies in EU CEE countries abound. The Trendchart database shows that around 30 per cent of all measures originate in the EU CEE countries.[25] The existing classification of measures does not take full account of the demand dimension.[26] There is one category out of five that captures some of the demand measures discussed above: 'markets and innovation culture'. Here, the two sub-categories of 'support for innovation culture' and 'support for the creation of new markets' are the most relevant, and indicate that the number of measures in those sub-categories is extremely low across all EU countries. In the EU CEE countries there is only one out of nine measures that is directed to support for the innovation culture, although demand begins with curiosity, with a positive attitude towards innovation and the abilities required to absorb it. Further, half of the EU CEE countries do not have any innovation policy measures in place to support the diffusion of innovation through fiscal means (and the other EU member states have only two measures).

A systematic scan of all the measures in the Trendchart Database reveals some interesting facts.[27] Within the category 'support for the creation of new markets', Poland has a set of measures within the Operational Programme that focus on the application of new technologies. For example, the so-called 'New investments with high innovation potential' programme supports new private investment covering the application of new highly innovative organizational and technological solutions in production and services, including those leading to a reduction in the adverse impact on the environment. The Polish 'Technology Credit Scheme' provides support for the implementation of new technology, with a strong focus on novelty assessed by independent experts. The Czech Republic marketing measure supports especially small and medium-sized enterprises (SME) in their international marketing efforts, for example, international trade fairs. Lithuania[28] has a 'Modernization of enterprises and implementation of Innovations' programme, which links the purchase of 'modern' equipment (especially information technology related) to consultancy to facilitate the take-up and application of innovative products, thus increasing business demand for innovation.

One area where there is more systematic support for modernizing through innovations is eco-efficiency. In the wake of the Lisbon Strategy, most countries have started to put great emphasis on eco-efficient technologies and their diffusion. For example, in Bulgaria, one of the activities within the Operational Programme is designed to improve industry through a financial incentive to include eco-efficient technologies in investment for modernization. Although the target is diffusion of existing leading edge eco-efficient technologies (not new to the market innovations), green procurement is receiving greater visibility than innovation procurement. It is too early to forecast outcomes, but this emphasis could enable a broader understanding of the options provided by an orientation towards demand.

There are a small number of measures in the category 'support for the creation of favourable innovation cultures'. For example, the InnoAwareness programme in Estonia sets out to inform the general public and companies about the importance of innovation and its diffusion through various publications and mobilization activities.

Information on public procurement in the area of innovation indicates that it is not part of the innovation policy tool box or a concern for decision-makers,[29] although in some parts of CEE awareness is improving, showing that the general European discussion is having some influence. The challenges in terms of public procurement related to inefficiencies and favouritism (see above) are obvious, and the building of efficient and transparent procurement structures and practices is a necessary step towards serious public procurement measures designed to spur innovation.

There are two areas where public procurement is linked to innovation, green procurement (linked to the Lisbon strategy goals) and eGovernment. Green procurement is part of most national reform programmes, but progress is slow and is oriented towards the diffusion of existing eco-efficient technologies. However, green procurement is one means to enlarge the rationale of public procurement, and by doing so, also opens up the space for bringing in innovation.[30] In eGovernment, innovations are linked to making public administrations more efficient. European intelligence (www.epractice.eu/) provides information on eGovernment initiatives in all EU countries. We can see that there is a link between improving public services and improving Internet and ICT diffusion and user demand for ICT products. For example, the Estonian Information Society Strategy 2013 claims that the 'public sector is a smart customer, ensuring that as much freedom as possible is left for innovative solutions as regards in public procurement' (IDABC, 2008: 9).

At the same time, it is necessary for there to be some balance between the latest innovative developments available, and the latest technologies, in order to increase their diffusion. Box 7.1 describes the ID transportation card in use in Tallinn, and shows how a range of services has been introduced on the back of this electronic ID card, and how this innovative solution has spread to another market. This example demonstrates some principles for public procurement of innovation (italicized in Box 7.1).

7.3.3.3 Long-term demand articulation

A prerequisite for long-term successful innovation policy based on demand is clear definition of (future) societal demands. Demand acts as a signal to the market and, as indicated above, in order to become meaningful in the marketplace, needs must be translated into demands expressed in and mediated through the marketplace (Mowery and Rosenberg, 1979).[31] One means to define long-term societal demands and articulate them in the marketplace is through foresight exercises.

The foresight exercises in the CEE countries provide evidence of many activities, some large-scale (Nyiri, 2002; Radosevic, 2002; Havas and Keenan, 2008). However, most of these foresight programmes concentrate on the supply side – even when they include wider institutional contexts and link the science and technology setting to societal goals. Most demand future direction for the science base and knowledge production more

BOX 7.1   ID-TICKET FOR THE PUBLIC
TRANSPORTATION SYSTEM, TALLINN*

The ID-ticket is an electronic ticket for the public transport system
(bus–tram–trolley), on sale to users via an electronic payments
collection system with proof of ownership of a personal identification document (national ID-card). It is sufficient to be carrying an
ID-ticket to use the public transportation system; the conductor
has a machine that checks validity. ID-tickets can be purchased
via Internet banks, mobile phones or at sales points. Estonia
started issuing national ID-cards (to which the e-.ticket links) in
January 2002 (*new functions, increased effectiveness*). Without
the existence of this infrastructure, several innovative public
services in Estonia (e.g. eVoting) would not be possible (*multi-
plier effect of new technological trajectory*). In addition to being
a physical identification document, the ID-card has advanced
electronic functions facilitating secure authentication and a legally
binding digital signature for public and private online services.
An electronic processor chip (a smart card reader is needed for
operation) contains a personal data file as well as a certificate
of authentication. Certification Center Ltd., the issuing organization, was established as a 100% privately owned company in
2001, and as of 2007 is the only certification authority providing
certificates for authentication and digital signing for Estonian
ID-cards (*public–private partnership*). The procurement process
generated bids from six applicants; one was a joint tender from
AS Certification Center, AS EMT (mobile telephone operator) and
AS Eesti Ühispank (bank), which was selected. The service was
introduced successfully in 2004. No fixed price was agreed. The
price was to be based on tickets sold: 4.49% of returns to go to
the procuring company (*intelligent incentive structures*). Thus,
the company had an interest in the application working efficiently.
Next to the small financial risk, the technological risk was small
and the technologies developed relied on an already existing ID-
card-based infrastructure (*novel combination an introduction to
new contexts rather than radical innovation*). The main concern
was whether the new service would be welcomed by users (*successful absorption*). It was not the intention for the ID-ticket to
replace the old channels of distribution, but to create additional
ones. It has proved a very efficient service, especially from the

standpoint of controlling the usage of tickets: (1) the ticket is personalized, so it is not possible to share it (which was possible with the old style transport tickets); (2) the process to control the validity of the tickets in public transport is quick *(increased effectiveness)*. The ID-ticket generated interest in obtaining a national ID-card. A similar service is now operating in Tartu, Estonia *(potential infant lead market)*. The intellectual property belongs to the certification centre. In sum, the ID-ticket is an example of a successful innovation where product, process and organizational innovations are combined.

Source: *Example and text author and adapted from Lember et al. (2007: 43–44).

generally, and seek guidance on priority-setting (see for example the Czech approach as described in Havas and Keenan 2008: 297–99).

There are very few examples of foresight processes that set out clear and specific demands and/or make this clear demand the starting point for future innovation and science and technology policy initiatives. One example of a demand-driven programme is eGovernment foresight in Bulgaria. This activity was conducted in the context of the European FORETECH project (http://foretech.online.bg). The project report speculates on expected effects. This demonstrates the link between satisfying immediate societal demands and triggering modernization and spillover effects in other areas.

Crafting e-government is a process with significant multiplication effects – it will improve quality of service delivery, reduce corruption and save time for citizens and businesses and, thus, will increase productivity and welfare. The side-effects could be a growing ICT industry, greater demand for information technology specialists and higher requirements for civil servants, and increased computer penetration. Building e-government foresight will probably result in building scenarios or even foresights in other tangential areas such as education or ICT industry (ARCF/CCIT, 2003: 6).

7.4   CONCLUSION AND POLICY RECOMMENDATIONS

This brief review of the multitude of innovation policy measures based on the EU Trendchart does not do justice to the various attempts undertaken and falls short of being a quantitative or qualitative assessment. Country-specific studies would be needed to assess activities more thoroughly. However, the overall finding is that in the CEE countries efforts to articulate and support demand for innovations, both public and private, are not keeping pace with attempts to boost supply (of science and technology) and compete in the international race for better supply conditions.

The reasons for this lack of demand orientation and slow pace of change are manifold. In section 7.3 we discussed the challenging demand conditions and proposed a set of policy challenges. Here we propose a set of recommendations for the EU CEE to tackle the challenges and design complementarities based on the demand side. It should be stressed that these recommendations are general; they take account of some context conditions common to the EU CEE as outlined above.

(1)   A Complementary Strategic Rationale

Demand-oriented innovation policy faces ideational challenges. New thinking needs time to diffuse and policy-making and administrations in the EU CEE – and in other countries – are locked into established analytical and policy models that are transferred across borders and often not questioned as to their suitability for the diverse contexts of the CEE countries. While the supply side, the build-up of research capabilities and related infrastructure are important, the dominance of the science-based, high-tech model in science and innovation policy discourse is counterproductive. Demand-oriented approaches must be able to counter resistance from established science communities that fear for the freedom of science even when demand-oriented innovation policies do not dictate research lines, but rather formulate problems and seek solutions and applications. Policies to support demand and the supply of technology do not have to be conflicting. As the Aho report (Aho et al., 2006: 47) rightly points out, the supply side, especially in cohesion regions of the EU, is best served if linked to the local context and needs. The CEE countries need strong policy discourse to construct their own theories.

Given the obvious challenges on the supply side in the EU CEE countries, demand-based innovation policy is more likely to face resistance if satisfying leading edge domestic demand means reliance on foreign suppliers which might not invest in the country (satisfying demand through imports). As we have shown, this is a serious issue for many of the EU CEE countries. However, demand policies can be selective – targeting areas where local producers are in a good position to compete and deliver, to encourage local involvement based on knowledge of local demand conditions. More importantly, and linked to the next recommendation, policies need to focus on absorption and diffusion to modernize industries rather than targeting national champions. They need to encourage the upgrading of specific industries, or encourage areas where foreign supply can be matched to local maintenance skills or to sub-contracting, and where foreign firms could establish production near to local demand. In other words, an initial lack of leading edge suppliers should not be a barrier. Benefits should accrue to society, directly and indirectly, from good demand-based policies.

(2) Three Clear Priorities in Innovation Policies

(a) Demand-oriented policies should try to influence the innovation culture in society in general, encourage more risk-taking among consumers (alongside firms and administrations) as buyers, and increase awareness of innovations and enable their use. This process starts with primary education and runs through the arsenal of education and training programmes;

(b) Innovation policy should focus on the link to the need to modernize industry, and upgrade equipment and organizational processes, and support the building of a business community that is willing and able to absorb technologies and organizational innovations. This should be a major focus of innovation policy;

(c) A better understanding is needed that policy oriented to innovation will be more powerful if linked to other societal goals. Policy support for innovation should be better defined in terms of meeting societal challenges. This would alter the role of innovation policy fundamentally, make innovation policy a moderator, facilitator and enabler for other ministries and administrations, and would link innovation efforts with demand articulated in other policy areas.

(3) Institutions and Governance Mechanisms Need to be Brought in Line with New Requirements

Current governance structures are not designed to provide a broader innovation policy approach. A new and much more committed horizontal coordination between ministries is needed to facilitate government to promote large-scale, demand-driven activities and set clear priorities. The facilitating and supporting role of innovation policy needs to be coordinated in new ways. In most CEE countries science and innovation policies (if coordinated at all) are mainly geared towards supply-side issues and often are not suited even to this traditional model.

(4)   A Long Journey Towards Public Procurement for Innovation

Administrations constitute a major institutional challenge for public procurement. The existence of favouritism and poor administrative capabilities adds complexity in many of the CEE countries. It cannot be denied that to establish an appropriate level of trust and to build credible structures that provide checks against unfair practices are crucial pre-requisites for procurement to become a meaningful mechanism to boost innovation in EU CEE countries. Innovation procurement requires flex-ibility, open formulation of functional specifications, flexibility in the bidding process and the application of the necessary complex negotiation or technical discourse procedures as outlined in the EU directives. These conditions are difficult for any country: the existence of favouritism will not help in the transition towards those improved practices. It impedes innovation and is a barrier to the opening of the system to innovative practices.

Beyond the obvious need to play by the rules, there are other govern-ance issues. Public procurement needs to be organized in ways that make procurement less risky, bring together the purchasing decision-maker, the administrative user and the procurer, and make the procurement process open to external users and to industry in general. Strategic procurement should be driven by those policy-makers who are responsible for certain domain policies and administrative mission and those knowledgeable about applications, user requirements and supplying markets.

This involves the skills to define needs in functional terms in order to enable industry to respond with innovations.

(5)   Making Good Use of European Initiatives and Support

Support at EU level, mainly in the form of the Structural Funds, should be monitored and checked against the principles outlined above. Priority should be given to instruments that demand or accelerate the diffusion of innovations and the spillovers identified above. There is a danger that the budget invested through the Structural Funds will not be used to its full potential.

The administrations in EU CEE countries should engage more widely and more intensively in practices and instruments that help them to learn. For example, the OMC-Nets appears to offer opportunities for EU CEE countries with specific structural challenges, to brainstorm about demand-oriented policies, specific market needs and policy instruments.[32] Procurement networks across the EU to implement the EU Lead Market initiative should be followed up actively, even if the supply-side economic benefits are not immediately obvious. These activities can act as a laboratory for independent initiatives to bundle the demands of a set of EU CEE countries to enable large-scale public procurement or new sectoral or infrastructure initiatives in specific regions.

(6) Appropriate Strategic Intelligence

Demand-oriented innovation policy needs to be underpinned by analytical evidence. The role of strategic intelligence in demand-based approaches is complex, and governments must build the capacities within and outside administrations to:

- formulate policy rationales and intervention logics (achieved through strategic intelligence on the supply side);
- understand and measure effects and support learning;
- understand bottlenecks on the demand side in order to decide about intervention;
- support technology and market intelligence – not to pick winners, but to understand how functional requirements potentially can be met by suppliers;
- organize a discourse between users and potential supplier firms;
- articulate demand, as the current fixation on technology foresight makes way for a more balanced approach to the definition of visions for the future of society and industry.[33]

(7) Communication of the Approach: Not Government Knowing it All

As was stated in the introduction to this chapter, demand-based policy is not a silver bullet and, as this overview shows, it poses severe challenges especially for the EU CEE countries – and challenges that differ among these countries. However, a bundle of principles and reforms as outlined above could open the way to a better understanding of what demand-based innovation policy is: it should serve the economy *and* society, and complement rather than substitute for supply-side policies. Mobilizing all government functions to make society and industry more keen to adopt innovation is not about building a technocratic machine to pick winners and steer markets, but is about organizing discourse in the best possible way and formulating strategies based on demand. Policy oriented to demand is more than just another instrument: it is related to the innovation culture, the readiness and willingness to ask for and adopt innovation among society, and improving the responsiveness of the system to the needs and the challenges in society and individual countries. To communicate these benefits clearly and to avoid the perception of a relentlessly intervening state is the most important first task.

NOTES

*   I am grateful to some of the contributors to this volume, and especially Slavo Radosevic, for very valuable comments on an early version of this chapter. The usual disclaimer applies: all shortcomings remain the sole responsibility of the author.

1. For a broader introduction see Edler (2007a, 2007b, 2007c, 2010), on which this chapter draws.

2. We should stress that this definition should not be confused with, or reduced to, demand-based approaches to *science* policy where priorities for funding in science are derived from societal or economic preferences; measures of support, mainly financial, continue to be geared towards the supply side or knowledge creation. This is in line with Bitzer and Hirschhausen's (1998) definition of demand-oriented science and technology policy for CEE countries.

3. These authors are referenced for purposes of illustration only and adopt quite different perspectives to the meaning of demand for market development. We return to their work below.

4. Although in their more narrow definition the authors do not classify all the variables mentioned as demand variables, they are clearly part of the overall demand conditions of an economy, since trust and corruption impinge directly on public procurement for innovation, and absorption of technology indicates the inclination and readiness of businesses to absorb new technologies.

5. One means to overcome these asymmetries is constructive technology assessment, in which societal actors not directly involved in the production of a technology are included in the technology discourse at an early stage.

6. Source: http://www.proinno-europe.eu/index.cfm?fuseaction=page.display&topicID= 275&parentID=51#.

7. For more detail see Radosevic (2004a: 648).

8. (1) Interest in science and technology; (2) optimism about science; (3) attitude towards the risks involved in new technologies; (4) attitude to the future; (5) attitude to the environment; (6) attitude to other cultures; and (7) customer responsiveness (inclination of customers to absorb innovation).

9. Excludes Romania and Bulgaria.

10. *The World Competitiveness Yearbook*, issued by UIMD Switzerland, is equally comprehensive in its country coverage; however, the data do not include demand conditions. See http://www.imd.ch/research/publications/wcy/Factors_and_criteria.cfm.

11. The sample size for responding EU CEE countries ranges from Slovenia (75 firms) to Lithuania (109 firms).

12. This variable is not independent of the supply of technology, but its definition highlights the will or ability of firms to demand and absorb new technologies (see Table 7.2).

13. This means that products are not purchased on the basis of their initial cost, but on the basis of their cost over the life cycle of the product, which includes running costs, maintenance, and so on.

14. See Transparency International 2008: http://www.transparency.org/news_room/ in_focus/2008/cpi2008/cpi_2008_table.

15. There is one striking deviation between the WEF and Transparency International data: Slovakia is rated higher in the latter.

16. This analysis does not include Romania and Bulgaria.

17. Italy and Greece are non-Eastern European (EE) outlier countries, which are ranked lower than the five EE countries.

18. The problem of insufficient domestic supply *and* FDI to meet rising demand is mentioned repeatedly in the analysis of innovation systems in EE; see, for example, Romania Trendchart (2007: 26) available at http://www.proinno-europe.eu/page/ innovation-and-innovation-policy-romania, accessed 13 August 2010.
19. For a country such as Bulgaria, with the lowest level of firm R&D in Europe (ERAWatch, 2008; Edler et al., 2008), this is a particularly problematic assessment.
20. http://www.proinno-europe.eu/index.cfm?fuseaction=page.display&topicID=261&par entID=52.
21. http://cordis.europa.eu/erawatch/.
22. The European Commission-funded Policy Mix Project has produced a framework to understand the interplay of RTDI policies: see http://www.policymix.eu/ PolicyMixTool/page.cfm?pageID=201.
23. See also the Structural Funds and innovation and knowledge dimension evaluations, http://ec.europa.eu/regional_policy/sources/docgener/evaluation/rado_en.htm.
24. See http://www.czechinvest.org/data/files/innovation-text-of-the-programme-412.pdf, p. 6.
25. It is thus not possible to match the innovation policy measures database to the catalogue of demand-based instruments, as outlined above.
26. A statistical analysis of the relative weight of demand-oriented measures based on the Trendchart is thus not feasible.
27. The scan shows that not all measures within this category support demand; some are R&D support measures.
28. http://www.proinno-europe.eu/index.cfm?fuseaction=wiw. measures&page=detail&id=9040.
29. Based on a set of interviews conducted in Bulgaria and Lithuania.
30. Interview with EU official responsible for green procurement in the Lisbon Strategy.
31. Mowery and Rosenberg (1979: 140) stress the distinction between 'need' and 'demand', and define the latter as a function of preferences and income, denoting a systematic relationship between price and quantity.
32. OMC-Nets: Open Method of Coordination networks in support of research policies.
33. For a discussion of the meaning of foresight for innovation policy strategy making and the role of demand therein, see also Georghiou and Harper (2008).

REFERENCES

Aho, E. et al. (2006), *Creating an Innovative Europe*, Report of the Independent Expert Group on R&D and Innovation appointed following the Hampton Court Summit, Chaired by Mr Esko Aho, Brussels: European Commission.

Anderson, E.S. (2007), 'Innovation and demand', in H. Hanusch and A. Pyka (eds), *Elgar Companion to Neo-Schumpeterian Economics*, Cheltenham, UK and Northampton, MA, USA: Edward Elgar, pp. 754–65.

ARCF/CCIT (Applied Research and Communications Fund/Coordination Centre on Information, Communication and Management Technologies) (2003), 'Background chapter: e-government foresight in Bulgaria', Sofia, April, available at http://foretech.online.bg/docs/Background%20chapter-e-gov-Bg. pdf, accessed December 2008.

Beise, M., J. Blazejczak, D. Edler, K. Jacob, M. Jänicke, T. Loew, U. Petschow, K. Rennings et al. (2003), *The Emergence of Lead Markets for Environmental Innovations*, Berlin: Institute for Ecological Economy Research (FFU).

Bhide, A. (2006), 'Venturesome consumption, innovation and globalization', paper prepared for the Joint Conference of CESifo and the Center on Capitalism and Society 'Perspectives on the Performance of the Continent's Economies', Venice, 21–22 July, available at http://www.bhide.net/bhide_venturesome_consumption.pdf, accessed December 2008.

Bitzer, J. and C. Hirschhausen (1998), 'Science and technology policy in Eastern Europe: a demand-oriented approach', *DIW Quarterly Journal of Economic Research*, **2**, 139–48.

Blind, K. (2007), 'Regulation in nachfrageorientierte Innovationspolitik' ('Regulation in demand oriented innovation policy'), in J. Edler (ed.), *Nachfrageorientierte Innovationspolitik. Benchmarking-Studie im Auftrag des Deutschen Bundestag (Demand-oriented Innovation Policy. Benchmarking Study on behalf of the German Parliament)*, Stuttgart: Fraunhofer IRB, pp. 277–89.

Bruno, N., M. Miedzinski, A. Reid and M.R. Yanzi (2008), 'Socio-cultural determinants of innovation', *WP 10 Innovation Watch Project*, available at http://www.technopolis-group.com/resources/downloads/WP10-Socio-cultural-factors.pdf, accessed December 2008.

Cohen, W. and D. Levinthal (1990), 'Absorptive capacity: a new perspective on learning and innovation', *Administrative Science Quarterly*, **35**(1), 128–52.

Dalpé, R. (1994), 'Effects of government procurement on industrial innovation', *Technology in Society*, **16**(1), 65–83.

Dalpé, R., C. DeBresson and H. Ciaoping (1992), 'The public sector as first user of innovations', *Research Policy*, **21**(3), 251–63.

David, P., P. Aghion and D. Foray (2008), 'Science, technology and innovation for economic growth: linking policy research and practice in "STIG Systems"', *MPRA Chapter No. 12096*, available at http://mpra.ub.uni-muenchen.de/12096/, accessed December 2008.

Eaton, J. and S. Kortum (1995), 'Engines of growth', *NBER Working Paper No 5207*, Washington, DC: National Bureau for Economic Research.

Ebersberger, B. (2007), 'Nachfrageorientierte Innovationspolitik in Finnland' ('Demand-oriented Innovation Policy in Finland'), in J. Edler (ed.), *Nachfrageorientierte Innovationspolitik. Benchmarking-Studie im Auftrag des Deutschen Bundestag (Demand-oriented Innovation Policy. Benchmarking Study for the German Parliament)*, Stuttgart: IRB, pp. 115–29.

Edler, J. (2007a), 'Nachfrageorientierte Innovationspolitik: eine konzeptionelle Einführung' ('Demand Oriented Innovation Policy: a conceptual introduction'), in J. Edler (ed.), *Bedürfnisse als Innovationsmotor – Konzepte und Instrumente nachfrageorientierter Innovationspolitik (Needs as Drivers of Innovation – Concepts and Instruments of Demand Oriented Innovation Policy)*, Berlin: Sigma, pp. 41–60.

Edler, J. (2007b), 'Demand based innovation policy', *Working Paper 9*, Manchester: Manchester Institute of Innovation Research, available at http://www.mbs.ac.uk/research/workingpapers/index.aspx?AuthorId=4464.

Edler, J. (ed.) (2007c), *Bedürfnisse als Innovationsmotor: Konzepte und Instrumente nachfrageorientierter Innovationspolitik (Needs as Drivers of Innovation: Concepts and Instruments of Demand Oriented Innovation Policy)*, Berlin: Sigma.

Edler, J. (2010), 'Demand based innovation policy', in R. Smits, S. Kuhlmann and P. Shapira (eds), *The Theory and Practice of Innovation Policy: An International Research Handbook*, Cheltenham, UK and Northampton, MA, USA: Edward Elgar.

Edler, J. and L. Georghiou (2007), 'Public procurement and innovation: Resurrecting the demand side', *Research Policy*, **36**, 949–63.

Edler, J., M. Beatson, A. Hodson, R. Kaarly and P. Koch (2008), 'Policy mix peer reviews: country report: Bulgaria', A Report of the CREST Policy Mix Expert Group, CREST – European Union Scientific and Technical Research Committee, available at http://ec.europa.eu/invest-in-research/pdf/download_en/pol_mix_bu.pdf, accessed 13 August 2010.

Edler, J., M. Beatson, S. Smits, B. Pukl and J. Windmüller (2007), *Policy Mix Peer Reviews: Country Report Lithuania*, Brussels: European Commission DG Research.

Edler, J., L. Hommen, M. Papadokou, J. Rigby, M. Rolfstam, L. Tsipouri and S. Ruhland (2005), Innovation and Public Procurement. Review of Issues at Stake, Study for the European Commission, Final Report, available at http://cordis.europa.eu/innovation-policy/studies/full_study.pdf, accessed 13 August 2010.

Edquist, C. (ed.) (1997), *Systems of Innovation. Technologies, Institutions and Organizations*, London and Washington, DC: Pinter Publishers.

Edquist, C., L. Hommen and L. Tsipuri (2000a), *Public Technology Procurement and Innovation*, Boston, MA: Kluwer Academic Publishers.

Edquist, C., L. Hommen and L. Tsipuri (2000b), 'Policy implications', in C. Edquist, L. Hommen and L. Tsipuri (eds), *Public Technology Procurement and Innovation*, Boston, MA: Kluwer Academic Publishers, pp. 301–11.

ERAWatch (2008), 'EW Country Report 2008: Bulgaria', EW Reports, CORDIS, available at http://cordis.europa.eu/erawatch/index.cfm?fuseaction=reports.content&topicID=1119&parentID=592, accessed 13 August 2010.

Fontana, R. and M. Guerzoni (2007), 'Incentives and uncertainty. An empirical analysis of the impact of demand on innovation', *SPRU Electronic Working Paper Series 163*, Brighton: Science and Technology Policy Research – SPRU, available at http://www.sussex.ac.uk/spru/documents/sewp163.pdf.

Gatignon, H. and T.S. Robertson (1985), 'A propositional inventory for new diffusion research', *Journal of Consumer Research*, **11**, 849–67.

Georghiou, L. (2007), 'Demanding innovation', NESTA Provocation 02; London/Manchester, available at http://www.nesta.org.uk/library/documents/demanding-innovation.pdf.

Georghiou, L. and J.C. Harper (2008), 'FTA for research and innovation policy and strategy', paper prepared for the Third International Seminar on Future-Oriented Technology Analysis: Impacts and Implications for Policy and Decision-making, Seville, 16–17 October.

Geroski, P.A. (1990), 'Procurement policy as a tool of industrial policy', *International Review of Applied Economics*, **4**(2), 182–98.

Havas, A. and M. Keenan (2008), 'Foresight in CEE Countries', in L. Georghiou, J. Harper, M. Keenan, I. Miles and R. Popper (eds), *The Handbook of Technology Foresight*, Cheltenham, UK and Northampton, MA, USA: Edward Elgar, pp. 287–318.

Hollanders, H. and A. Arundel (2007), 'Differences in socio-economic conditions and regulatory environment: explaining variations in national innovation performance and policy implications', INNO-Metrics Thematic Report, available at www.proinno-europe.eu/admin/uploaded_documents/eis_2007_Socio-economic_conditions.pdf, accessed March 2010.

IDABC (2008), eGovernment Factsheet Estonia, October 2008, available at http://www.epractice.eu/en/document/288215, accessed 13 August 2010.

Jacob, K. and M. Jänicke (2003), 'Lead märkte für umweltinnovationen' ('Lead Markets for Environmental Innovations'), *Politische Ökologie*, **21**(84), 19–21.

Lember, V., T. Kalvet, R. Kattel, C. Penna and M. Suurna (2007), 'Public procurement for innovation in Baltic metropolises', Tallinn, available at http://www.baltmet.org/uploads/filedir/File/BMI-K-75%20Study%20Public%20Procurement%20for%20Innovation%20in%20Baltic%20Metropolises%20(Final)%202%200%20Lember%20(2).pdf, accessed December 2008.

Lundvall, B.-Å. (ed.) (1992), *National Systems of Innovation. Towards a Theory of Innovation and Interactive Learning*, London: Pinter Publishers.

McMeekin, A., K. Green, M. Tomlinson and V. Walsh (eds) (2002), *Innovation by Demand: An Interdisciplinary Approach to the Study of Demand and its Role in Innovation*, Manchester and New York: Manchester University Press.

Meyer-Krahmer, F. (2004), 'Vorreiter-Märkte und Innovation. Ein neuer Ansatz der Technologie und Innovationspolitik' ('Lead markets and innovation. A new approach to technology and innovation policy'), in F.W. Steinmeier and M. Machnig (eds), 'Made in Germany '21. Innovationen für eine gerechte Zukunft' ('Made in Germany. Innovation for a fairer future'), Hamburg: Hoffman und Campe, pp. 95–110.

Ministry of Economy and Energy (2007), *Annual Report on the Bulgarian National Innovation Policy*, Sofia: Ministry of Economy and Energy.

Mowery, D. and N. Rosenberg (1979), 'The influence of market demand upon innovation: a critical review of some recent empirical studies', *Research Policy*, 8(2), 102–53.

Neji, L. (1998), 'Evaluation of Swedish market transformation programmes', paper prepared for the ACEE Summer Study on energy efficiency in buildings, Panel II.

Neji, L. (1999), 'Evaluation of Swedish market transformation programmes', paper prepared for the Conference of the European Council for an Energy Efficient Economy, Summer Study Proceedings 1999.

Nyiri, L. (2002), 'How to turn "Mobilising regional foresight potential" into a structural contribution to integration', paper prepared for Conference 'Europe's Regions Shaping the Future: the Role of Foresight', Brussels, 24–25 September, available at ftp://ftp.cordis.europa.eu/pub/foresight/docs/9-howtoturn.pdf, accessed 13 August 2010.

Polt, W., P. Koch, B. Pukl and B. Wolters (2007), 'OMC policy mix review report: Country report: Estonia', October, available at http://ec.europa.eu/invest-in-research/pdf/download_en/omc_ee_review_report.pdf, accessed December 2008.

Porter, M.E. (1990), *The Competitive Advantage of Nations*, New York: The Free Press.

Radosevic, S. (2002), 'Regional policy, national and regional foresight in Central and Eastern European candidate countries', paper presented at 'Europe's regions shaping the future: the role of foresight', ETAN Conference on 'Mobilising the regional foresight potential for an enlarged EU', July, available at ftp://ftp.cordis.europa.eu/pub/foresight/docs/9-howtoturn.pdf, accessed March 2010.

Radosevic, S. (2004a), 'A two-tier or multi-tier Europe? Assessing the innovation capacities of Central and East European countries in the enlarged EU', *Journal of Common Market Studies*, 42(3), 641–66.

Radosevic, S. (2004b), '(Mis)match between demand and supply for technology: innovation, R&D and growth issues in Central and Eastern Europe', in W.L. Filho (ed.), *Supporting the Development of R&D and the Innovation Potential of Post-Socialist Countries*, Amsterdam: IOS Press, pp. 71–82.

Rothwell, R. and W. Zegveld (1981), 'Government regulations and innovation: industrial innovation and public policy', in R. Rothwell and W. Zegveld (eds), *Industrial Innovation and Public Policy*, Westport, CT: Greenwood Press, pp. 116–47.

Schmookler, J. (1962), 'Economic sources of inventive activity', *Journal of Economic History*, **2**(1), 1–20.

Smits, R. (2002), 'Innovation studies in the 21st century: questions from a user's perspective', *Technological Forecasting and Social Change*, **69**(9), 861–83.

Suvilehto, H.-M. and E. Överholm (1998), 'Swedish procurement and market activities: different design solutions for different markets', Proceedings of the 1998 ACEEE (American Council for Energy Efficient Economy) Summer Study on Energy Efficiency in Buildings.

von Hippel, E. (1986), 'Lead users, "A Source of Novel Product Concepts"', *Management Science*, **32**(7), 791–805.

Wilkinson, R., L. Georghiou, J. Cave, C. Bosch, Y. Caloghirou, S. Corvers, R. Dalpé, J. Edler, K. Hornbanger, M. Mabile, M.J. Montejo, H. Nilsson, R. O'Leary, G. Piga, P. Tronslin and E. Ward (2005), 'Procurement for Research and Innovation. Report of an Expert Group on measures and actions to assist in the development of procurement practices favourable to private investment in R&D and innovation', EC, DG Research, EUR 21793 EN, Brussels, available at: http://ec.europa.eu/invest-in-research/pdf/download_en/edited_report_18112005_on_public_procurement_for_research_and_innovation.pdf, accessed 13 August 2010.

World Bank (2007), 'Bulgaria. Accelerating Bulgaria's convergence: the challenge of raising productivity', Volume II, Main Report, Report 38570, Washington, DC: The World Bank.

World Economic Forum (WEF) (2008), *Global Competitiveness Report 2008–2009*, Davos: World Economic Forum, available at http://www.weforum.org/en/initiatives/gcp/Global%20Competitiveness%20Report/PastReports/index.htm, accessed March 2010.

8. Innovation policy options for 'catching up' by the EU CEE member states

Philippe Aghion, Andreas Reinstaller,
Fabian Unterlass, Jakob Edler,
Anna Kaderabkova, Rajneesh Narula,
Slavo Radosevic and Alasdair Reid

8.1 INTRODUCTION

The INCOM Workshop that forms the basis for this volume was organized as a personal initiative of the Czech Prime Minister, Mr Mirek Topolánek, within the Czech EU presidency. The objective of the workshop was to re-examine EU innovation policies, especially those of the new member states (NMS) from Central and Eastern Europe (CEE). The focus on these countries stems from an understanding of the limitations of a 'one size fits all' approach to innovation policy and a realization that some measures may not be effective for countries at a greater distance from the world technology frontier.

Some national innovation systems (NIS) are capable of generating disruptive innovations, while others focus mainly on incremental innovation. Countries operating at the technology frontier have the capacity to generate innovations that are new to the world, while those behind the frontier are capable mainly of only imitative innovation. Innovative imitation should not be underestimated, however, since it can be the basis for major gains in wealth and technological competency.

In view of the differences in innovation capacities among the EU27 it is unrealistic to expect that similar policies and indicators can be used to gauge and benchmark the innovation performance of such diverse membership. The diversity of the EU27 brings benefits, but exploiting them calls for new policy approaches.

Based on this reasoning, the workshop gathered a large number of participants from academia, consultancies, industry and the policy-making community including the European Commission (EC) and the Organisation for Economic Cooperation and Development (OECD). Expert papers were presented and their policy implications discussed in a series of round-table sessions. Contributions to this volume, which are based largely on papers submitted to the workshop and the discussions, form the basis for the actions proposed in this chapter.

8.2   KEY ISSUES

We highlight several issues in calling for the formulation of country-specific 'policy mixes'.

8.2.1   Governance Failure

There is widespread failure in governance related to innovation policy in the CEE NMS and the 'old' EU15. Contributors to this volume as well as workshop participants noted that the coordinating and funding mechanisms and processes that apply to the EU27 often constrain the ability of member states to adapt appropriately to meet new growth challenges and achieve strategic policy objectives. The EU's renewed Lisbon Strategy[1] and its re-examination of mechanisms such as the open method of coordination (OMC), which are important for establishing a European Research and Innovation Area, recognize this challenge.

The action points proposed in this chapter are an attempt to reconsider the current approaches to innovation policy. Members of the expert group who contributed to this volume and who prepared the analytical material for the INCOM Workshop are cognizant of the need to take the cross-country heterogeneity within the enlarged EU and among existing institutions and technological requirements as the starting point to an adaptive innovation strategy. This perspective needs similarly to be acknowledged and focused on in constructing policies for innovation and growth. It points to the need for a larger measure of autonomy and greater differentiation in policy prescriptions.

8.2.2   The Global Financial Crisis

The second issue is related to the treatment and policy responses to the global financial crisis (GFC) 2008–09, which has provided the opportunity to develop new approaches to pro-growth innovation policies. Fiscal stimulus packages that address macroeconomic stabilization could also fund innovation programmes aimed in the short term at boosting innovation and, over time, increasing competitiveness. Fiscal policy could provide a way to balance the need for short-term stimulus and the long-term benefits of sustainable growth with the need for greater autonomy and differentiation in policy prescriptions.

8.2.3 The EU27: Diversity in Innovation Capacities vs Similarity in Innovation Policies

The third issue concerns the gap between the widely varying innovation capacities of the EU27 countries on the one hand, and the apparent similarity in the innovation policies being applied to this group of countries on the other. The EU27 includes innovation leaders, innovation followers, moderate innovators and catching-up countries (cf. European Innovation Scoreboard – EIS). The evolution of these groups of countries is driven by different factors and, while some of the challenges apply to all of the EU27, each country faces unique problems related to maintaining competitiveness through innovation.

8.3   EUROPE'S CHANGING INNOVATION LANDSCAPE

Some of the drivers of innovation-based growth, such as education, training, skills acquisition and research and development (R&D), are common to all the EU27 countries. However, accession to the EU of the CEE NMS in 2004 and 2006 changed the EU's socioeconomic landscape considerably and produced much wider variation in the levels of development across EU members. The EU27 encompasses some of the most technologically advanced economies in the world as well as countries that are still in the catching-up stage and whose growth potential lies in their endowment of relatively low-cost labour, good education systems, geographic location and general infrastructure.

8.3.1 Differences in the Drivers of Innovation

This volume presents evidence that, despite pockets of R&D excellence, innovation in the CEE NMS and some Southern European countries is driven less by R&D-based innovation than by technology transfer and non-technological modes of innovation. For many enterprises in these countries, the main challenge is to develop in-house capacities (through training, adoption of new management practices, and recruitment of qualified staff), and to absorb and adapt new technologies that allow them to compete in international markets. There is a need for incentives to encourage the adoption of technologies and innovation as well as the development of new in-house capabilities. Foreign direct investment (FDI) plays a crucial role in augmenting skills, improving management practices and introducing new technologies in national economies. Reputable international firms construct the foundations for future generations of national entrepreneurs through training, and the promotion and provision of vital hands-on experience to employees.

8.4   EU POLICIES AND BENCHMARKS PREDATING NMS ACCESSION

The main lines of EU-level innovation policy were conceived before accession of the NMS to the Union. The original Lisbon Agenda was established in 2000; it was revised in 2005, but this revision did not account sufficiently for the problems and needs of NMS, and conflicts with the requirement that the basic characteristics of each country (for example level of technological development, financial constraints, nature of its institutions) should be catered to. The Europe 2020 Strategy adopted in summer 2010 similarly sets uniform policy objectives and benchmarks across the EU. The EU27 represents a new architecture of countries, no longer characterized by a simple East–West divide. Rather, the majority of innovation indicators suggest that there is a four-tier Europe based on innovation capacities (see Figure 8.1).

Benchmarks, such as the EIS, measure all EU economies in terms of the level of growth based on innovation – and especially world frontier technology, but not against a set of criteria tailored to their individual technology profiles. It would be difficult to construct a composite benchmark that would capture both innovation and imitation activities, and direct and indirect R&D (that is R&D embodied in imported and domestic inputs and capital goods), and which would be equally relevant across all the EU27 countries.

8.5   THE NEED FOR A VARIETY OF BENCHMARKS AND TAILORED ANALYSES

As a starting point, we need a variety of appropriate benchmarks that can be exploited by national policy-makers to position their countries against peers and to monitor trends over time. More in-depth evidence-based analysis of national, regional and sectoral innovation performance

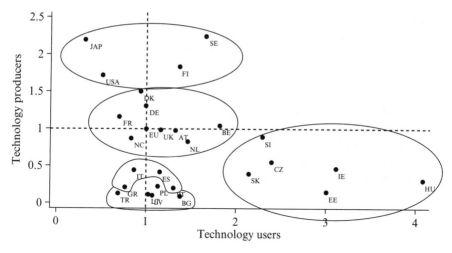

Notes:
Includes domestic investment in R&D, and domestic embodied technology as well as direct and indirect R&D embodied in foreign technology. GDP per capita in purchasing power parities are used to capture the state of economic development in each country.
Clusters were identified using average linkage and squared Euclidian distance measure.

Source: M. Böheim, A. Reinstaller and F. Unterlass (2009), 'Sectoral innovation modes and innovation policy: implications for the New Member States', paper presented at the Innovation for Competitiveness INCOM Prague, 22–23 January and OECD ANBERD, EUROSTAT/OECD Input Output tables. WIFO calculations based on NIFU-STEP calculations.

Figure 8.1   Country clusters: technology users and technology producers, based on the direct and indirect domestic and foreign share of R&D in GDP

is needed to understand specific situations and to inform policy debate in a range of diverse countries. Experience from the CEE NMS, for example Estonia, where the public authorities have funded a range of policy studies, suggests that this investment leads to better designed policy instruments, which better address the specific demands and capacities of local firms.

8.6   NEED FOR COUNTRY-SPECIFIC POLICY MIXES

In attempting to become the world's most competitive economy, the EU is targeting investment in innovation and knowledge to generate faster growth and increased rates of employment. For individual member states, this translates into sets of specific requirements; however, common to all is that growth requires policies that are complementary, that is, coherent innovation and competition policies, and higher education, labour market and financial market policies that are aligned. This complementarity is essential for an innovation policy to be effective, that is, able to overcome any weaknesses in national framework conditions and policies.

The mix of innovation and complementary policies will inevitably be country-specific and must reflect each country's traditions, aspirations and institutional needs, and distance from the technology frontier. For example, competition policy will play different roles in different countries, and the education policies of different member states will be focused on different types of skills. Most countries require support for the formulation and implementation of well-conceived, early-stage finance programmes adapted to local conditions and needs, a requirement that has become especially acute since the global economic crisis and the disappearance of many sources of early-stage finance.

8.7   RETHINKING THE EU INNOVATION AGENDA

The INCOM workshop adopted a perspective that takes account of cross-country heterogeneity in NIS and differences in innovation capacity, and which recognizes the need for country-specific policy mixes and complementarities between innovation policy, structural reform and other policies.

It acknowledged the need for a reconsideration of existing policies and a rethinking of the EU innovation agenda. For example, although EU policies allow for some flexibility in implementing the Lisbon Agenda, many countries are focusing on improving a few R&D-based indicators (for example, the Barcelona goal of 3 per cent of GDP spent on R&D) at the expense of other measures that could foster national innovation capability and competitiveness. These Lisbon Agenda and the Barcelona goals create strong public and peer pressure to improve certain indicators, which could be detrimental for countries where formal R&D is not the main innovation input for industry and is of less relevance for their catching-up process.

8.8   POLICY RECOMMENDATIONS

There is some evidence that the EU Lisbon Agenda has failed to deliver: most member states have not implemented strategies that are coherent with their industry specializations or levels of economic development. In order to work towards Europe 2020, the successor to the Lisbon Agenda, we need an approach that addresses this contextual variety. In some countries the challenges and policy mixes will be the same, but not in all. The important issues and differences will become even more significant in the context of the 2020 strategy, which will focus especially on articulating and implementing EU governance systems and policy strategies, in areas considered important for growth and employment. We can expect more EU-wide and compulsory strategies '*to pro-actively help national policies to converge*, build on European synergies and raise everyone's performance and standards of excellence'(emphasis added), in the future.[2]

Based on the evidence presented, there are several actions that might be considered by policy-makers.

8.8.1 Reconsidering the Balance for Funding Between Excellence and R&D Relevance

R&D funding systems in the NMS have made significant progress towards openness and a focus on excellence. Access to the EC Framework Programmes and other EU R&D funding initiatives has stimulated the drive towards world excellence. While the funding of high-level research is important, it must be industry- and socially relevant. Achieving this balance has been a cause for concern in terms of ensuring that innovation will be relevant to the economies and societies of the NMS.

The INCOM workshop in 2008 was in complete agreement that research capacities must be linked to firm-level demand through co-funding mechanisms, involving the business enterprise sector, which would help to connect pockets of scientific excellence with the business sector. The workshop participants believed that co-funding would enhance the impact of science parks, high-tech incubators and similar initiatives. Building the capacities for applied research in the public sector, and technology absorption and technology development in close cooperation with commercial enterprises, should enhance the competitiveness of national firms.

8.8.2 Improving Education

High quality education and development, in particular, of cognitive skills, are essential for building the absorptive and innovation capacity of the NMS. The OECD PISA (Programme for International Student Assessment) results indicate that the level of education in the majority of NMS is far from satisfactory. There is a need for stricter teaching standards, and staff appraisal at all levels in the system. Education systems should support creativity by increasing levels of specialization in late school careers, and should abandon the culture of rote learning.

It will be important for the NMS to reconcile the quality and equity of education provision. However, higher quality is unlikely without increased expenditure. Most NMS spend less per student than the OECD average. Improvements are required at all levels of education.

8.8.3   A Focus on Training and Lifelong Learning

The problem of skills mismatches in the labour market is not confined to the NMS, but is a major structural problem for these countries. Investment in lifelong learning and retraining to help workers and firms to upgrade their skills continuously is not a priority in the NMS countries, but needs to become more prominent. There is evidence that providing tax incentives for workers and firms to take up training opportunities is more effective than attempts to set up publicly-managed training programmes. The GFC should not be used as an excuse to reduce funding for education and training.

8.8.4   Evaluating the Use and Design of Structural Funds to Improve Innovation Capacity

In many NMS, Structural Funds (SF) often substitute for national funding for innovation. There are strict limits to the substitutive role of SF, and there is an urgent need to re-evaluate their role as an innovation policy instrument, and their effectiveness as an incentive for long-term, sustainable achievement. The bureaucracy involved in exploiting SF is often heightened by the rules of national administrations and creates barriers for small firms and universities. A re-evaluation is needed of the role of SF in building firm- and industry-specific infrastructures with strong linkages to public research systems. There should be a greater focus on searching for opportunities to attract or enhance the impact of additional FDI from the perspective of growth rather than redistribution.

8.8.5   Exploring Institutional Approaches for Integrating FDI and Innovation Policy

The role of FDI in technology transfer has significance for the NMS. Foreign investment has been important for accelerating the pace of economic growth and creating industry and export specializations. As the cost-based competitiveness of the NMS is eroded, and the transition to knowledge-based economies becomes more urgent, the role of foreign investors will be central. The challenge will be to attract and motivate innovation-intensive FDI inflows and re-investment. The NMS will need to integrate FDI and innovation policy both organizationally and financially, with the objective of orienting the EU SF to building industry- and technology-specific support services. This will require foreign investment agencies embedding or integrating foreign investors in the local economy and both developing existing FDI and attracting new investment. The embedding of investment should be seen as a continuous activity to integrate FDI in local economies to stimulate multiple linkages by opening up the opportunities for industrial upgrading by both domestic and foreign firms.

8.8.6   Spurring Innovation Through Demand Policies

One of the obstacles to the NMS becoming part of the knowledge-based economy is their limited local markets and weak demand for R&D and innovation. Enhancing demand for innovative goods and services and, therefore, R&D, will be more difficult than initiating supply-side R&D measures (investment in research infrastructure, and so on).

A vision that includes incentives: The key to linking innovation to demand is to include innovation policy within the broader national strategy. For example, incentives for industrial modernization should be linked to increased demand for the latest production technologies and related services. Public programmes could search for and establish as standard the most efficient models in the market and support for industry investment could be linked to these standards.

Public procurement: Public procurement has been shown to stimulate demand for innovation in Japan and the US. However, it is not a feature of any of the NMS and its introduction there to enhance innovation would face several challenges. These include lack of appropriate incentive structures and skills in public administration systems, widespread cronyism in some countries, exclusion of foreign organizations from public procurement tenders and the building of national champions, and involvement of small and medium sized enterprises in public procurement contracts. These problems apply in varying degrees to all European countries and could be overcome through systematic and strategic actions. The ideal situation would be one in which public needs were served, and industry was incentivized to deliver innovation and to upgrade services and maintenance activities.

8.8.7 Investing in Strategic Intelligence and Administrative Capacity for Improved Policy-making

The capacities of the NMS countries to formulate, evaluate and implement innovation policy appropriate to their development and to technological and institutional contexts are insufficient. There is a strong need to improve policy evaluation, policy analysis and policy-making capabilities within and outside of government (cf. non-governmental organizations). This calls for systematic training at postgraduate (masters) level, short training courses in innovation, R&D management, policy, and evaluation to produce larger numbers of qualified personnel able to deal with innovation policy issues in government ministries.

8.8.8 Initiating EU calls for Research to Develop Differentiated Approaches to Innovation and Growth Policy

There is an overall low level of strategic intelligence on the drivers of innovation and growth in the NMS. The EC, through its Framework Programmes, should promote more social science research in the area of innovation, growth and competitiveness. There is an especial need for support for research that focuses on innovation policy issues relevant to specific groups of EU countries, including the NMS group. For example, the EC Competitiveness and Innovation Programme could provide more opportunities for stakeholders from specific groups of countries to work together on embedding FDI and linking it to innovation policy.

An EU-level initiative should be launched to develop additional indicators that will capture differences in industry specialization, institutions, economic and social development and governance structures. These indicators and analyses would form the basis for evidence-based policy-making in the member states.

NOTES

1. Commission of the European Communities (2005), *Working Together for Growth and Jobs: A New Start for the Lisbon Strategy*, Communication to the Spring European Council, Communication from President Barroso in agreement with Vice-President Verheugen, COM (2005) 24, http://ec.europa.eu/growthandjobs/pdf/COM2005_024_en.pdf.
2. A. Mettler (2010), *If Not Now, Then When? Using Europe 2020 to Restore Confidence and Growth*, The Lisbon Council, available at http://www.lisboncouncil.net/publication/publication/59-ifnotnowthenwhen.html.

Index